S0-BFP-031

Medical and Biologic Effects of Environmental Pollutants

CARBON MONOXIDE

Committee on
Medical and Biologic Effects of
Environmental Pollutants

DIVISION OF MEDICAL SCIENCES
ASSEMBLY OF LIFE SCIENCES
NATIONAL RESEARCH COUNCIL

NATIONAL ACADEMY
OF SCIENCES
WASHINGTON, D.C. 1977

Other volumes in the Medical and Biologic Effects of Environmental Pollutants series (formerly named Biologic Effects of Atmospheric Pollutants):

ARSENIC (ISBN 0-309-02604-0)
ASBESTOS (ISBN 0-309-01927-3)
CHLORINE AND HYDROGEN CHLORIDE (ISBN 0-309-02519-2)
CHROMIUM (ISBN 0-309-02217-7)
COPPER (ISBN 0-309-02536-2)
FLUORIDES (ISBN 0-309-01922-2)
LEAD (ISBN 0-309-01941-9)
MANGANESE (ISBN 0-309-02143-X)
NICKEL (ISBN 0-309-02314-9)
NITROGEN OXIDES (ISBN 0-309-02615-6)
OZONE AND OTHER PHOTOCHEMICAL OXIDANTS (ISBN 0-309-02531-1)
PARTICULATE POLYCYCLIC ORGANIC MATTER (ISBN 0-309-02027-1)
SELENIUM (ISBN 0-309-02503-6)
VANADIUM (ISBN 0-309-02218-5)
VAPOR-PHASE ORGANIC POLLUTANTS (ISBN 0-309-02441-2)

NOTICE: The project that is the subject of this report was approved by the Governing Board of the National Research Council, whose members are drawn from the Councils of the National Academy of Sciences, the National Academy of Engineering, and the Institute of Medicine. The members of the Committee responsible for the report were chosen for their special competences and with regard for appropriate balance.

This report has been reviewed by a group other than the authors according to procedures approved by a Report Review Committee consisting of members of the National Academy of Sciences, the National Academy of Engineering, and the Institute of Medicine.

The work on which this publication is based was performed pursuant to Contract No. 68-02-1226 with the Environmental Protection Agency.

Library of Congress Cataloging in Publication Data

National Research Council. Committee on Medical and Biologic Effects of Environmental Pollutants.
Carbon monoxide.

(Medical and biologic effects of environmental pollutants)
Bibliography: p.
Includes index.
1. Carbon monoxide—Toxicology. 2. Carbon monoxide—Environmental aspects.
3. Air—Pollution. I. Title.
RA577.C36N37 1977 615.9'25'681 77-12771
ISBN 0-309-02631-8

Available from:
Printing and Publishing Office, National Academy of Sciences, 2101 Constitution Ave., N.W., Washington, D.C. 20418

Printed in the United States of America

SUBCOMMITTEE ON CARBON MONOXIDE

RONALD F. COBURN, University of Pennsylvania School of Medicine, Philadelphia, Pennsylvania, *Chairman*

ERIC R. ALLEN, State University of New York Atmospheric Sciences Research Center, Albany, New York

STEPHEN M. AYRES, St. Louis University School of Medicine, St. Louis, Missouri

DONALD BARTLETT, JR., Dartmouth Medical School, Hanover, New Hampshire

EDWARD F. FERRAND, New York City Department of Air Resources, New York, New York

A. CLYDE HILL, University of Utah, Salt Lake City, Utah

STEVEN M. HORVATH, University of California Institute of Environmental Stress, Santa Barbara, California

LEWIS H. KULLER, University of Pittsburgh Graduate School of Public Health, Pittsburgh, Pennsylvania

VICTOR G. LATIES, University of Rochester School of Medicine and Dentistry, Rochester, New York

LAWRENCE D. LONGO, Loma Linda University School of Medicine, Loma Linda. California

EDWARD P. RADFORD, JR., The Johns Hopkins University School of Hygiene and Public Health, Baltimore, Maryland

JAMES A. FRAZIER, National Research Council, Washington, D.C., *Staff Officer*

COMMITTEE ON MEDICAL AND BIOLOGIC EFFECTS OF ENVIRONMENTAL POLLUTANTS

Acknowledgments

Members of the Subcommittee on Carbon Monoxide, under the chairmanship of Dr. Ronald F. Coburn, wrote this report. It is appropriate to mention here that the same people also wrote the section on carbon monoxide in the report for the U.S. Senate Committee on Public Works.*

Drs. Coburn and Eric R. Allen wrote the Introduction. Dr. Allen also wrote Chapter 2, on properties and reactions, and Chapter 3, on sources, occurrence, and fate of atmospheric carbon monoxide. Dr. Edward F. Ferrand wrote Chapter 4, on environmental analysis and monitoring, and Appendix A, on methods of monitoring.

For Chapter 5, Dr. Edward P. Radford, Jr., wrote the material on uptake; Drs. Coburn and Donald Bartlett, Jr., on physiologic effects; Dr. Lawrence D. Longo, on effects on the pregnant woman, the developing embryo, the fetus, and the newborn infant; Dr. Lewis H. Kuller and Dr. Stephen M. Ayres, on cardiovascular effects; Dr. Victor G. Laties, on behavioral effects; Dr. Steven M. Horvath, on effects during exercise and effects on populations especially susceptible to carbon monoxide

*National Academy of Sciences, National Academy of Engineering, Coordinating Committee on Air Quality Studies, *Air Quality and Automobile Emission Control*, vol. 2, *Health Effects of Air Pollutants*, U.S. Senate Committee Print Serial No. 93-24 (Washington, D.C.: U.S. Government Printing Office, 1974), 511 pp.

exposure owing to reduced oxygenation at altitudes above sea level; Dr. Bartlett, on the effects of chronic or repeated exposure; and Dr. Coburn, on dose-response characteristics in man.

Dr. A. Clyde Hill wrote Chapter 6, on the effects of carbon monoxide on bacteria and plants.

In Chapters 7 and 8, the statements in the summary and the recommendations were written by the members of the Subcommittee and assembled by Dr. Coburn.

Dr. Radford wrote Appendix B, on measurement in biologic samples.

The Environmental Protection Agency's Air Pollution Technical Information Center supplied information from its computer data base. Dr. Robert J. M. Horton of the EPA obtained various technical and scientific documents and other resource information. The staff of the NRC Assembly of Life Sciences Advisory Center on Toxicology gave assistance in obtaining resource information and also reviewed the report.

The report was reviewed by the Academy's Report Review Committee with the assistance of anonymous reviewers selected by that Committee. The members of the MBEEP Committee reviewed the report in depth. It was also reviewed by the Associate Editor, Dr. Henry Kamin, and five anonymous reviewers selected by him.

The staff officer for the Subcommittee on Carbon Monoxide was Mr. James A. Frazier. We also acknowledge the staff support of Mrs. Renée Ford for editing the report, Ms. Joan Stokes for preparation and verification of the references, and Mr. Norman Grossblatt for his editorial assistance.

Contents

1	Introduction	1
2	Properties and Reactions of Carbon Monoxide	4
3	Sources, Occurrence, and Fate of Atmospheric Carbon Monoxide	28
4	Environmental Analysis and Monitoring	46
5	Effects on Man and Animals	68
6	Effects on Bacteria and Plants	168
7	Summary and Conclusions	172
8	Recommendations	186
	Appendix A: Methods of Monitoring Carbon Monoxide	191
	Appendix B: Measurement of Carbon Monoxide in Biologic Samples	199
	References	205
	Index	233

1

Introduction

Man has experienced the effects of carbon monoxide poisoning at least since that period in prehistory when he first discovered the art of making and utilizing fire. Numerous accounts of tragic events, circumstances, and phenomena that can be directly or indirectly attributed to the toxic properties of carbon monoxide have been related in folklore and mythology. Lewin,[258] who traced the early history of carbon monoxide, was led to the conclusion that this form of poisoning is unique in its close association with the history of civilization. For example, carbon monoxide poisonings drastically increased in the fifteenth century when the use of coal for domestic heating increased. These poisonings were attributed both to inhalation of carbon monoxide formed by incomplete combustion in the heating of homes and to the exposure of coal miners to the deadly "white damp," encountered after underground explosions and mine fires.

The introduction of illuminating gas (a mixture of hydrogen, carbon monoxide, methane, and other hydrocarbons, also known as carbureted water gas) for domestic heating purposes further increased the hazard of carbon monoxide poisoning. Although this fuel is still used extensively in Europe, it has largely been replaced in the United States with natural gas. More recently, the introduction of the internal-combustion engine and the development of numerous technological processes in which carbon monoxide is produced have increased still further the hazard of exposure to this toxic gas. Despite awareness for many centuries that human exposure to combustion fumes was hazardous, it was not until 1919 that industrial production of carbon monoxide was recognized to be an environmental health problem of national importance. At the First International Congress of Labor, the rapidly increasing use of the internal-combustion engine as a source of industrial power was cited as a major contributor of the carbon monoxide being inhaled by workers. The exhaust from the internal-combustion engine is the principal contemporary anthropogenic source of carbon monoxide. Over the centuries, the problem of dealing with carbon monoxide exposure has expanded from dwellings to work environments and now includes the ambient air in cities. This report reflects our concern with the adverse effects of exposure to carbon monoxide at the concentrations found in our urban and industrial air.

At the time that the Air Quality Criteria for Carbon Monoxide appeared in 1970, data were produced that suggested that, when man received an acute carbon monoxide exposure that produced carboxyhemoglobinemia as low as 3% saturation, there were adverse effects on complex mental functions such as vigilance. There was also epidemiologic evidence relating the incidence of myocardial infarction and the concentration of carbon monoxide in air. The possibility was raised that a significant fraction of the urban population might be experiencing adverse health effects due to carbon monoxide. A major criticism was that it was not known whether the data could be extrapolated to urban population groups.

Studies on the biologic effects of carbon monoxide on man since 1971 apparently support the conclusion that carboxyhemoglobin levels as low as 3–5% may have effects on vigilance and aerobic metabolism under conditions of exercise at maximal oxygen uptake. It is uncertain whether these experimental results can be extrapolated to urban populations, but it is strongly suspected that they can. In addition, an awareness of a spectrum of carbon monoxide susceptible populations is growing, particularly in patients with coronary vascular diseases and in the fetus.

This document summarizes the carbon monoxide literature related to effects on man and his environment for the consideration of the Environmental Protection Agency in updating the information in the Air Quality Criteria for Carbon Monoxide. It emphasizes recent major advances in our knowledge of carbon monoxide: chemical reactions in air; biologic effects on man; problems in monitoring urban concentrations and relating such data to the exposure of populations; data concerning the identification of susceptible populations; and evidence implicating carbon monoxide as a causal factor in disease. We have not tried to review all published articles but only those deemed to be the important studies related to carbon monoxide air quality criteria. There is a large literature on adverse effects of cigarette smoking, and some of these effects may be related to carbon monoxide.

2

Properties and Reactions of Carbon Monoxide

Carbon monoxide (CO) is an imperceptible poisonous gas, the most common source of which is the incomplete combustion of carbonaceous materials. It is probably the most publicized and the best known of all air quality criteria pollutants* because of the frequency of accidental deaths attributed to its inhalation over the years. Its toxic and sometimes lethal properties, however, are due to acute effects resulting from exposures to very high concentrations in confined spaces, generally exceeding 500 ppm for several hours. In this report we are mainly concerned with the deleterious effects resulting from human exposures to much lower concentrations over considerably longer periods of time. In the latter context we are also concerned with the role this oxide of carbon plays as a chemically reactive environmental pollutant and atmospheric trace constituent. This requires a detailed quantitative understanding of major physical and chemical factors governing production, control, transformation, and removal of carbon monoxide prior to and during its atmospheric life cycle. Recently, there has been a renewal of interest in reactions involving carbon oxides because of their fundamental importance in the genesis and evolution of planetary atmospheres as well

*Those pollutants for which an air quality criteria document has been published as required by the Clean Air Act.

as in pollutant-forming combustion, flame and explosion processes, and terrestrial atmospheric chemistry.

In this report the fundamental scientific and technical knowledge concerning carbon monoxide in its role as an air pollutant affecting public health and welfare is updated and assessed. Special emphasis is placed on the more recent advances in our understanding and relevant discoveries made during the last 5 yr.

For a comprehensive review of pertinent literature published prior to 1972, there are two excellent reports, one commissioned by the National Air Pollution Control Administration, U.S. Public Health Service,[450] and the other by the North Atlantic Treaty Organization, Committee on Challenges of Modern Society,[331] published in March 1970 and June 1972, respectively. They treat the topic of air quality criteria for carbon monoxide in considerable detail and provide a basis for this updated review. In addition, Cooper[109] has compiled an extensive carbon monoxide bibliography with abstracts of the literature published prior to 1966 that provides a further source of information.

GENERAL PHYSICAL AND CHEMICAL PROPERTIES

Carbon monoxide, a stable compound, is a heteropolar diatomic molecule. It absorbs radiation in the infrared region corresponding to the vibrational excitation of the electronic ground state of the molecule,[192] CO ($X^1\Sigma_g^+$). Because thermal population of excited vibrational levels is extremely inefficient, even at 1,000 C less than 1% of all the molecules present will reside in the first vibrationally excited ($\nu' = 1$) state above the ground vibrational state ($\nu'' = 0$). Radiation in the visible and near-ultraviolet regions of the electromagnetic spectrum is not absorbed by carbon monoxide, but in the vacuum ultraviolet region, a structured relatively weak absorption band extends from 155 to 125 nm. In this so-called spectroscopic fourth positive band system of carbon monoxide the electronic transition CO($A_\pi{}^1 \leftarrow X^1\Sigma_g^+$) occurs.[192,197] The characteristics of the absorption spectrum and the frequency-dependent extinction coefficients, as well as the photochemistry of carbon monoxide in the vacuum ultraviolet region, have been reviewed and discussed elsewhere.[67,305] Interest in carbon monoxide photochemistry has recently been stimulated by the discovery of this gas at about 0.1% in the Martian atmosphere[29,223] and by detection in the atmosphere of Venus.[108,386] These discoveries suggest that it plays a significant role in the development of primitive planetary atmospheres in our solar system.

Carbon monoxide has a low electric dipole moment (0.10 debye), short interatomic distance (1.13 Å), and high heat of formation from atoms, or bond strength (1,072 kJ/mol), suggesting that this diamagnetic molecule is a resonance hybrid of three structures:[344]

$$\text{a)}\ :\overset{+}{C}:\overset{-}{\underset{\cdot\cdot}{\ddot{O}}}: \qquad \text{b)}\ :C::\ddot{O}: \qquad \text{c)}\ :\overset{-}{C}:::\overset{+}{O}:$$

These structures correspond to forms with single, double, and triple covalent bonds, i.e.,

$$\overset{+}{C}-\overset{-}{O},\ C=O \text{ and } \overset{-}{C}\equiv\overset{+}{O}.$$

They all apparently contribute about equally to the normal state of the molecule, thus counterbalancing the opposing effects of the number of covalent bonds and charge separation. Carbon monoxide is isoelectronic with molecular nitrogen (N_2), the nitrosyl cation (NO^+), and the cyanide anion (CN^-). The similarity to nitrogen causes difficulties in its physical separation and identification in air. It is a colorless, odorless, and tasteless gas that is slightly lighter than air and difficult to liquefy. Although an anhydride of formic acid (HCOOH), it is unreactive with water and is only slightly soluble. General physical properties of carbon monoxide are presented in Table 2-1.

PRODUCTION AND PREPARATION

Carbon monoxide is produced when carbon or combustible carbonaceous material is burned in a limited supply of air or oxygen, i.e., under fuel-rich conditions.

It is manufactured on a large scale by reducing carbon dioxide with carbon at high temperatures. Below 800 C the reduction is slow, but above 1,000 C the conversion, corresponding to the endothermic Reaction 1, is quite fast and efficient:

$$C(s) + CO_2(g) \rightarrow 2CO(g) - 163\ \text{kJ/mol}. \qquad (1)$$

The monoxide is a major constituent of the synthetic fuels "producer gas" (25% CO), and "water gas" (40% CO). The former, used mainly for heating purposes, consists primarily of nitrogen and carbon monoxide and is prepared by passing air through a bed of incandescent coke, the residue of coal remaining after destructive distillation. If coal is used in place of coke, coal gas will also be present. This fuel mixture, commonly used for domestic heating in Europe, has been almost completely replaced by natural gas in the United States. "Water gas" is made when steam is blown through incandescent coke at 1,100 C. It is a mixture of hydrogen (49%) and carbon monoxide (44%):

TABLE 2-1 Physical Properties of Carbon Monoxide[467]

Molecular weight	28.01
Critical point	−140 C at 34.5 atm
Melting point	−199 C
Boiling point	−191.5 C
Density	
at 0 C, 1 atm	1.250 g/l
at 25 C, 1 atm	1.145 g/l
Specific gravity relative to air	0.967
Solubility in water[a]	
at 0 C	3.54 ml/100 ml (44.3 ppmm)[b]
at 20 C	2.32 ml/100 ml (29.0 ppmm)
at 25 C	2.14 ml/100 ml (26.8 ppmm)
Explosive limits in air	12.5–74.2%
Fundamental vibrational transition $CO(X'\Sigma_g^+, \nu' = 1 \leftarrow \nu'' 0)$	$2{,}143.3\ cm^{-1}$ ($4.67\ \mu m$)
Conversion factors	
at 0 C, 1 atm	$1\ mg/m^3 = 0.800$ ppm[c] $1\ ppm = 1.250\ mg/m^3$
at 25 C, 1 atm	$1\ mg/m^3 = 0.873$ ppm $1\ ppm = 1.145\ mg/m^3$

[a]Volume of carbon monoxide is at 0 C, 1 atm (atmospheric pressure at sea level = 760 torr).
[b]Parts per million by mass (ppmm = mg/l).
[c]Parts per million by volume.

$$H_2O(g) + C(s) \rightarrow CO(g) + H_2(g), \quad (2)$$

with traces of nitrogen (4%), carbon dioxide (2.7%), and methane (0.3%).

"Semi-water gas" is produced by blowing a mixture of steam and air through incandescent coke. Carbureted (enriched) water gas, which burns with a luminous flame, is prepared by mixing water gas with partially unsaturated hydrocarbons. "Water gas" burns with a blue, nonluminous flame, produces considerable heat in combustion, and may be used with Welsback mantles for illuminating purposes. The calorific values (available fuel energy) for combustion of producer and semi-water gas are low, of the order 125 Btu/ft^3 ($4.4 \times 10^6\ J/m^3$), compared with 350 Btu/ft^3 ($12.3 \times 10^6\ J/m^3$) for water gas and 600 Btu/ft^3 ($21 \times 10^6\ J/m^3$) for coal gas. Carbon monoxide can also be produced by several other methods, including:

- Reduction of carbon dioxide with zinc dust or iron filings at red heat—for example, with zinc dust:

$$CO_2(g) + Zn(s) \rightarrow CO(g) + ZnO(s) \quad (3)$$

• Heating charcoal with either zinc, iron, or manganese oxides—for example, with zinc:

$$C(s) + ZnO(s) \rightarrow CO(g) + Zn(s) \quad (4)$$

• Heating carbon with certain alkaline earth carbonates, such as chalk or barium carbonate—for example, with barium carbonate:

$$BaCO_3(s) + C(s) \rightarrow BaO(s) + 2CO(g) \quad (5)$$

On the laboratory scale, carbon monoxide for analytical and other purposes can be conveniently prepared by the following methods:

• The reaction of concentrated sulfuric acid (H_2SO_4) with formic acid at 100 C (Reaction 6), or oxalic acid at 50 C (Reaction 7), or sodium formate (Reaction 8) or potassium ferrocyanide at room temperature (Reaction 9):

$$HCOOH \xrightarrow[100\ C]{H_2SO_4} H_2O + CO(g) \quad (6)$$

$$(COOH)_2 \xrightarrow[50\ C]{H_2SO_4} H_2O + CO(g) + CO_2(g) \quad (7)$$

$$2HCOONa + H_2SO_4 \rightarrow Na_2SO_4 + 2H_2 + 2CO(g) \quad (8)$$

$$K_4Fe(CN)_6 + 6H_2SO_4 + 6H_2O \xrightarrow[25\ C]{} 2K_2SO_4 + FeSO_4 + 3(NH_4)_2SO_4 \quad (9)$$

In the reaction with oxalic acid (Reaction 7), note that equal volumes of carbon dioxide and carbon monoxide are produced. Some sulfur dioxide (SO_2) is generated in all the reactions (Reactions 6-9) by the reduction of sulfuric acid as shown in Reaction 10:

$$H_2SO_4 + CO(g) \rightarrow H_2O + SO_2(g) + CO_2(g) \quad (10)$$

• The reaction of chlorosulfonic acid with formic acid also produces carbon monoxide (Reaction 11):

$$HCOOH + CSO_3Cl \rightarrow H_2SO_4 + HCl + CO(g) \quad (11)$$

In the above procedures the carbon monoxide produced can be purified by bubbling through caustic soda and then drying over phosphorus pentoxide.

- Thermal decomposition of nickel carbonyl at 200 C yields a high-purity product:

$$Ni(CO)_4(g) \rightarrow Ni(s) + 4CO(g) \quad (12)$$

Carbon dioxide absorbs solar ultraviolet radiation in its first two absorption bands between 170 and 120 nm. At wavelengths less than 165 nm, sufficient energy is available to dissociate the dioxide into the monoxide and electronically excited atomic oxygen[60] according to the spin-conserving reaction:

$$CO_2(X^1\Sigma_g^+) + h\nu \rightarrow CO(X^1\Sigma_g^+) + O(^1D) \quad (13)$$

Here $CO_2(X^1\Sigma_g^+)$ is the electronic ground state of carbon dioxide, CO $(X^1\Sigma_g^+)$ is the electronic ground state of carbon monoxide, and $O(^1D)$ is an electronically excited state of atomic oxygen. The quantum efficiency (number of molecules of product per absorbed photon of radiation) of this process is almost unity near 150 nm.

Carbon monoxide is also a product of the pyrolysis or photolysis of many oxygenated organic compounds. For example, at wavelengths below 340 nm, aliphatic aldehydes, which are constituents of photochemical smog, may photolyze directly into hydrocarbon and carbon monoxide as shown in Reaction 14a:

$$RCHO + h\nu \rightarrow RH + CO \quad (14a)$$

In the case of formaldehyde (HCHO), the quantum efficiency of this process is about 0.5 at 313 nm and ambient temperatures, but the efficiency is much less for higher aliphatic aldehydes. The photolysis of aldehydes also produces acyl radicals (*R*CO) according to Reaction 14b:

$$RCHO + h\nu \rightarrow RCO + H \quad (14b)$$

In the absence of molecular oxygen, acyl radicals can spontaneously decompose into hydrocarbon radicals (*R*) and carbon monoxide. However, in air this reaction is suppressed by the dominant formation of acylperoxy radical (RCO_3) adducts with oxygen.

CHEMICAL REACTIONS OF CARBON MONOXIDE

Decomposition

Carbon monoxide is quite stable and chemically inert under normal conditions (25 C and 1 atm) despite a carbon valency of two. At higher temperatures it becomes reactive, behaves as though unsaturated, and can act as a powerful reducing agent—a property used in many metallurgical processes, such as in blast furnaces.

At high temperature, in the presence of a catalyst (palladium, iron, or nickel), carbon monoxide undergoes the reversible disproportionation reaction:

$$2CO \rightleftarrows C + CO_2 + 162\ kJ/mol \quad (15)$$

As the temperature is increased, the equilibrium fraction of carbon dioxide decreases (for example, the equilibrium percentages by volume of dioxide are 90 at 550 C, 50 at 675 C, 5 at 900 C, and less than 1 at 1,000 C). The reverse reaction is utilized industrially to manufacture the monoxide (see Reaction 1).

Photolysis of carbon monoxide at 129.5 nm (xenon radiation) results in disproportionation into carbon dioxide and carbon suboxide (C_3O_2).[138] Since the absorption of radiation at wavelengths <111 nm is required to photodissociate carbon monoxide into its component atoms, it has been postulated that the photochemical reaction involves electronically excited molecules, $CO(A^1\pi)$, and the following sequence of reactions has been suggested:

$$CO + h\nu \rightarrow CO(A^1\pi) \quad (16)$$

$$CO(A^1\pi) + CO \rightarrow CO_2 + C \quad (17)$$

$$C + CO \rightarrow C_2O \quad (18)$$

$$C_2O + CO \rightarrow C_3O_2 \quad (19)$$

At this wavelength (129.5 nm), the quantum efficiency of carbon monoxide removal is 0.8 ± 0.4, but at 147 or 123.6 nm it is almost zero.[176] Similar products are formed by electrical and high-frequency discharges in carbon monoxide.

COMBUSTION[343]

Carbon monoxide burns in air or oxygen with a bright blue flame but does not itself support combustion (see Reaction 20).

$$2CO + O_2 \rightarrow 2CO_2 + 565 \text{ kJ/mol} \quad (20)$$

Although heat is evolved during combustion, the fuel value is low (320 Btu/ft^3 or 11.3 × 10^6 J/m^3). The blue flames seen on top of a clear fire consist of burning carbon monoxide. It is presumably produced in fires due to the reduction of carbon dioxide that is formed in the lower regions of glowing fuel near the entering air draught as it rises through an incandescent mass of carbon (see Reaction 21). The carbon monoxide burns on top of the fire where an excess of air is available. An interesting feature of these processes is that the reaction of carbon with oxygen at temperatures producing carbon dioxide generates heat (exothermic process—see Reaction 21), whereas the reaction of carbon dioxide with carbon at the same temperature absorbs heat (endothermic process—see Reaction 11).

$$C(s) + O_2(g) \rightarrow CO_2(g) + 1{,}109 \text{ kJ/mol} \quad (21)$$

Thus, it is possible to adjust the ratio of air (or oxygen) to carbon dioxide so that a desired temperature can be maintained continuously. In industry, large quantities of carbon monoxide are formed during the incomplete combustion of charcoal or coke, and to a lesser extent by other carbonaceous fuels, in a limited supply of air, i.e., under fuel-rich conditions. The presence of carbon monoxide in furnace gases is used as an indicator of improper air supply and its estimation in flue gases is used as a check on the operating efficiency of the furnace.

Stoichiometric mixtures (2:1 by volume) of carbon monoxide and oxygen explode upon ignition in the presence of trace amounts of moisture or hydrogenous compounds, such as methane (CH_4) or hydrogen sulfide (H_2S). These necessary impurities apparently act as a source of oxyhydrogen free radicals (hydroxyl and hydroperoxyl), which provide low-energy reaction paths leading to an explosive chain branching mechanism. Explosive limits in oxygen range from 15.5 to 93.9% by volume of carbon monoxide. These proportions may be compared with the limits in air given in Table 2-1.

HETEROGENEOUS REACTIONS

Carbon monoxide is the anhydride of formic acid but it does not react with water (liquid or vapor) at room temperature. However, the forward reaction in the equilibrium with formic acid vapor shown by Reaction 22 may be achieved by the application of an electrical or high-frequency discharge in stoichiometric mixtures (1:1) of carbon monoxide and water vapor.

$$CO(g) + H_2O(g) \rightleftarrows HCOOH(g) \qquad (22)$$

The reverse dehydration process is accomplished by the catalytic action of metallic rhodium. Carbon monoxide at 120 C and 3 to 4 atm pressure is rapidly and completely absorbed by a concentrated solution of caustic soda forming sodium formate. This salt is also produced when carbon monoxide is passed over caustic soda or soda lime heated to 200 C (see Reaction 23).

$$NaOH(s) + CO(g) \rightarrow HCOONa(s) \qquad (23)$$

Anhydrous formic acid is manufactured economically in quantity from sodium formate by distillation with concentrated sulfuric acid.

In the presence of a suitable catalyst, such as freshly reduced metallic nickel, carbon monoxide can be hydrogenated to methanol (CH_3OH), methane (CH_4), or other organic compounds depending upon the conditions selected. Its conversion to methane is the basis for the flame ionization gas chromatographic separation procedure used for the detection and estimation of carbon monoxide in ambient air.

When carbon monoxide is passed over heated metal oxides such as lead oxide (PbO) in Reaction 24, oxygen is extracted leaving the metal.

$$PbO(s) + CO(g) \rightarrow CO_2(g) + Pb(s) \qquad (24)$$

Rapid and quantitative reaction with various oxides at high temperatures has been used for the estimation of carbon monoxide. Some examples are:

- with cuprous oxide at room temperature or copper-cupric oxide at 270–300 C, carbon dioxide produced is quantitatively absorbed by caustic soda solution, and the volumetric loss on absorption is measured in the Orsat–Lunge apparatus.
- with iodine pentoxide at 90 C, iodine vapor is formed stoichiometrically by Reaction 25 and the iodine is collected and estimated by iodometry.

$$I_2O_5(s) + 5CO(g) \rightarrow I_2(g) + 5CO_2(g) \quad (25)$$

• with mercuric oxide at 150 C, mercury vapor is released, which is measured spectrophotometrically at 254 nm.

$$HgO(s) + CO(g) \rightarrow Hg(g) + CO_2(s) \quad (26)$$

The rate of oxidation of carbon monoxide in oxygen, although insignificant in homogeneous mixtures, is known to be enhanced by metallic catalysts such as palladium on silica gel or a mixture of manganese and copper oxides (Hopcalite).[490] It has also been shown that carbon monoxide is made catalytically active by adsorption on hot metallic surfaces.[446] The reactions of nitrous oxide (N_2O) with carbon monoxide chemisorbed on copper, charcoal, Pyrex glass, or quartz surfaces at high temperatures (above 300 C) have been studied extensively[240,286,414,426] and found to be quite efficient in oxidizing carbon monoxide (see Reaction 27).

$$CO_{abs} + N_2O(g) \rightarrow CO_2(g) + N_2(g) \quad (27)$$

CARBONYL AND COORDINATION COMPOUNDS[316,405]

In the presence of light, equal volumes of carbon monoxide and individual halogens (F_2, Cl_2, Br_2, I_2) or cyanogen (C_2N_2) react to form the corresponding volatile and highly reactive carbonyl halides or cyanide. In Reaction 28:

$$CO + X_2 \rightarrow COX_2 \quad (28)$$

X is a univalent element. Carbonyl chloride (phosgene), formed by reaction with chlorine (Cl_2), is the best known of these compounds and is highly poisonous at a concentration that has been suggested to be considerably less than that for carbon monoxide. It is used frequently as the solvent in nonaqueous systems of acids and bases. Phosgene undergoes ammonolysis when passed into a solution of ammonia in toluene:

$$COCl_2 + 4NH_3 \rightarrow CO(NH_2)_2 + 2NH_4Cl \quad (29)$$

forming urea [$CO(NH_2)_2$] and the salt ammonium chloride, which provides an effective means of removing the phosgene. It is also used in the manufacture of urea. Carbonyl fluoride (COF_2) is extremely reactive, attacking glass and many metals.

When carbon monoxide is passed over heated sulfur or selenium, carbonyl sulfide (COS) and selenide (COSe), respectively, are produced. All of the carbonyl compounds mentioned above are used widely and frequently in industrial organic syntheses and organic chemical production.

Carbon monoxide forms volatile metallic carbonyl compounds in which the carbon monoxide molecule exclusively is attached to a single metal atom or group of metal atoms. These include the carbonyls of chromium, molybdenum, and tungsten in Group VI of the periodic table of the elements; rhenium in Group VII; iron, ruthenium, and osmium in Group VIIIA; cobalt, rhodium, and iridium in Group VIIIB; and nickel in Group VIIIC. Although these are coordination compounds, their unusual composition is determined more by their tendency to form closed (filled) electronic shells rather than by the valence of the central metal. For example, in simple binary carbonyls, containing one metal atom (M) in the complex, i.e., ($M[CO]_y$), the carbon monoxide donates two electrons to the shell; thus, the effective atomic number (EAN) of the metal M becomes the atomic number of the next higher inert gas element for a stable complex (atomic number of element + $2y$). Thus, the EAN's for some of the metals mentioned above are: 36 in chromium hexacarbonyl [$Cr(CO)_6$], iron pentacarbonyl [$Fe(CO)_5$], and nickel tetracarbonyl [$Ni(CO)_4$]; 54 in molybdenum hexacarbonyl [$Mo(CO)_6$] and ruthenium pentacarbonyl [$Ru(CO)_5$]; and 86 in tungsten hexacarbonyl [$W(CO)_6$] and osmium pentacarbonyl [$Os(CO)_5$]. Similar considerations apply to the complex carbonyls with several metal atoms or other co-coordinated groups.

The addition of carbon monoxide to metallic elements, as shown in Reaction 30 for the case of nickel tetracarbonyl, occurs with a large decrease in volume, and the formation is promoted by employing high pressures (750 atm).

$$Ni(s) + 4CO(g) \rightarrow Ni(CO)_4(l) \qquad (30)$$

In the large-scale commercial production of iron pentacarbonyl, reduced iron is heated at 180 to 200 C under 50 to 200 atm of carbon monoxide. Cobalt, molybdenum, and tungsten carbonyls are prepared on a commercial scale under similar conditions with sulfur as a promotor. These carbonyl compounds are used in the separation of metallic elements as a source of highly purified metals and in the manufacture of plastics. Nickel carbonyl is as highly poisonous as phosgene at comparable concentrations (1 ppm); these toxicity levels are well below that for carbon monoxide, and these compounds should, therefore, be handled with great care. At elevated temperatures (>200 C) the gaseous binary metallic carbonyls decompose into carbon monoxide and deposit a high-purity metallic element. Similarly, iron and nickel carbonyls in the gas phase or in solution can dissociate readily at room temperature when irradiated at 366 nm and shorter wavelengths.[332]

Carbon monoxide can penetrate heated iron and escape through the iron flues of stoves or furnaces that are operated with an insufficient supply of air. At high pressures, carbon monoxide reacts with solid iron and invariably iron pentacarbonyl is formed in gas cylinders containing compressed carbon monoxide or gases such as commercial hydrogen that contain carbon monoxide. This impurity can lead to erroneous results in reaction studies and in other experiments where cylinder gases are used.[52]

Other complex carbonyl derivatives that are prepared by partial substitution of metallic carbonyls are: carbonyl halides such as with iron ($Fe[CO]_4hal_2$); amines of the carbonyls and carbonyl halides, such as with rhenium ($Re[CO]_3py_2$, $Re[CO]_3py_2hal$); carbonyl hydrides such as iron carbonyl hydrides ($H_2Fe[CO]_4$ or $Fe[CO_2][CO \cdot H]_2$) and their metallic derivatives; mixed carbonyl-nitrosyl compounds such as iron nitrosyl carbonyl ($Fe[CO]_2[NO_2]$); and compounds with only one carbon monoxide group on the central metal such as potassium nickel cyanocarbonyl ($K_2[Ni(CN)_3CO]$). Mixed carbonyl halides are easily prepared by Hiebeh's method, which involves heating metallic halides with carbon monoxide at high pressure.

The aqueous solubility of carbon monoxide is enhanced by the formation of coordination compounds with metal atoms, e.g., copper (Cu), silver (Ag), gold (Au), and mercury (Hg). Owing to this property, the gas can be quantitatively absorbed by the following solutions:

- Hydrochloric acid, aqueous ammonia, or a potassium chloride solution of cuprous chloride that produces colorless crystals with the composition $CuCl \cdot CO \cdot H_2O$ and the probable structure:

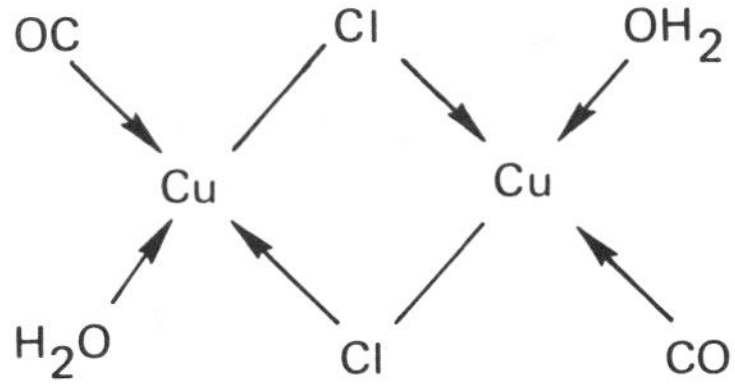

OC Cl OH_2

Cu Cu

H_2O Cl CO

- Silver sulfate solution in concentrated (fuming) sulfuric acid.
- *Dry* aurous chloride, forming benzene-soluble $Cl—Au \leftarrow CO$.

• Mercuric acetate in methanol; the product is convertible to the chloride complex by adding KCl:

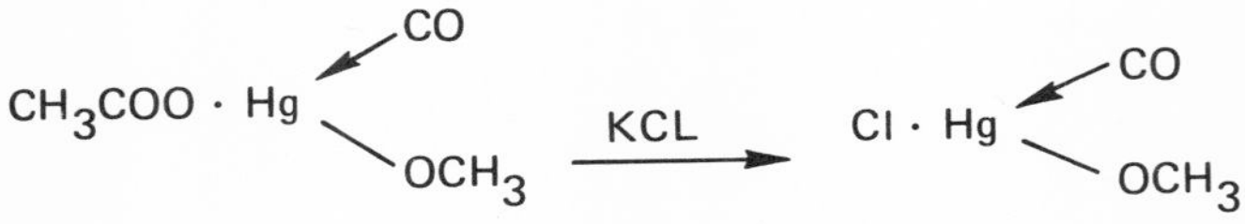

$$CH_3COO \cdot Hg \begin{matrix} \swarrow CO \\ \searrow OCH_3 \end{matrix} \xrightarrow{KCL} Cl \cdot Hg \begin{matrix} \swarrow CO \\ \searrow OCH_3 \end{matrix}$$

These coordination compounds can be used to estimate carbon monoxide selectively or to separate it from other gaseous mixtures.

The toxicity of carbon monoxide is due to its strong coordination bond formed with the iron atom in heme, which is 200 times stronger than that for molecular oxygen. The molecule ($C_{34}H_{32}N_4O_4Fe$) is a ferrous ion complex of protoporphyrin IX that, when combined with the protein globin, constitutes hemoglobin. Hemoglobin is made up of four subunits each with a heme moiety conjugated to a poplypeptide or globin, corresponding to 95% (by weight) globin. The planar, macro ring structure of heme is shown below.[490]

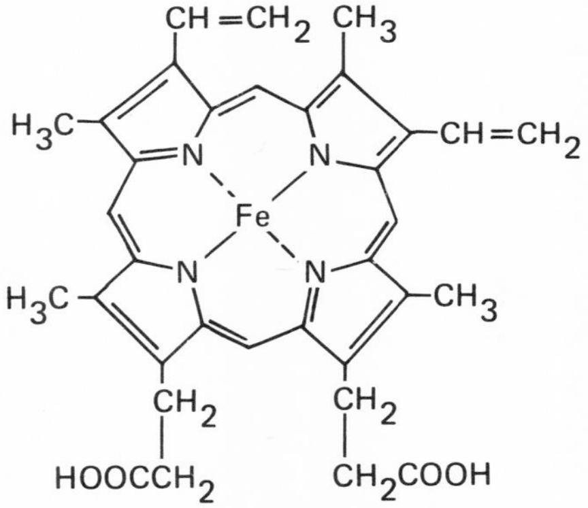

The purple compound carboxyhemoglobin, formed by absorption of carbon monoxide in blood, has a characteristic visible and near-ultraviolet absorption spectrum, which has been used in the determination of the amount of carbon monoxide uptake in humans and animals.

REACTIONS WITH ATMOSPHERIC CONSTITUENTS AND TRACE CONTAMINANTS

The efficiencies of gaseous atmospheric reactions capable of oxidizing carbon monoxide to carbon dioxide were reviewed by Bates and Witherspoon[31] in 1952.

Reactions with Molecular Species The homogeneous gas-phase reaction with molecular oxygen (O_2) (Reaction 32) is far too slow to be significant, even in urban atmospheres where carbon monoxide concentrations are relatively high (10–100 ppm).

$$2CO + O_2 \rightarrow 2CO_2 + 565 \text{ kJ/mol} \quad (32)$$

Experimental confirmation has been provided by the observed long-term stability (7 yr) of carbon monoxide in either dry or moist mixtures with oxygen (O_2) when exposed to sunlight.[306] Other possible homogeneous reactions with atmospheric constituents and trace contaminants include those with water vapor (H_2O), ozone (O_3), and nitrogen dioxide (NO_2).

$$CO + O_2 \rightarrow CO_2 + O + 33.5 \text{ kJ/mol} \quad (33)$$

$$CO + H_2O \rightarrow CO_2 + H_2 + 41.9 \text{ kJ/mol} \quad (34)$$

$$CO + O_3 \rightarrow CO_2 + O_2 + 423 \text{ kJ/mol} \quad (35)$$

$$CO + NO_2 \rightarrow CO_2 + NO + 226 \text{ kJ/mol} \quad (36)$$

The reactions with molecular oxygen (Reaction 33) and water vapor (Reaction 34) may occur in the lower atmosphere but are very slow and have high energy barriers (activation energies) to the reactions (213 and 234 kJ/mol, respectively). Also, at ambient temperatures (25 C) they exhibit very low molecular reaction collision efficiencies[139,171,447] ($< 10^{-15}$). Reactions 35 and 36, with ozone and nitrogen dioxide, also have been shown to have high activation energies[55,150,187,357,504] (84 and 117 kJ/mol, respectively). Therefore, the rates of Reactions 33 through 36 become significant only at substantially high temperatures (above 500 C) and consequently are unimportant under atmospheric conditions and at ambient concentrations at altitudes below 100 km. Above this level in the thermosphere, molecular kinetic temperatures increase rapidly with altitude[104] from 200 K (−73 C) at 100 km to 1,403 K (1,130 C) at 200 km and, thus, Reaction 33 can become significant. The increased thermodynamic probability that this reaction takes place above 100 km is at least partially offset by decreasing atmospheric densities ($< 3 \times 10^{-7}$ atm) and lower molecular collision frequencies ($< 2 \times 10^{3}\ s^{-1}$). At these high altitudes, competing reactions with atmospheric ions and electrons plus the solar photodissociation of carbon monoxide are possible.

Gas-Phase Reactions with Unstable Intermediates There is a possibility that rapid reactions can occur between carbon monoxide and certain reactive intermediates, such as atoms or free radicals, generated by chemical processes in the natural and polluted atmosphere. For example, oxygen atoms are produced by the photolyses of nitrogen dioxide ($\lambda <$ 436 nm) and/or ozone ($\lambda <$ 1,140 nm) in the sunlight near the ground, and in the upper stratosphere by the solar photodissociation of molecular oxygen at wavelengths in the range of 246 to 176 nm (corresponding to the Herzberg and Schumann-Runge absorption bands). Note that significant concentrations (ca. 0.02 ppm) of nitrogen dioxide and ozone are produced and play important roles in photochemical smog resulting from atmospheric chemical transformation in automobile exhaust products.

The probability of reaction between carbon monoxide and atmospheric atomic oxygen has been discussed by Leighton.[252] Estimates of the rate of Reaction 37, in which oxygen atoms [$O(^3P)$] and carbon monoxide molecules [$CO(X^1\Sigma_g^+)$]—both in their electronic ground state and accompanied by a chaperone molecule, M (to remove the excess energy)—react to form carbon dioxide [$CO_2(X^1\Sigma_g^+)$] in its electronic ground state, show that this spin-forbidden process is insignificant in air.

$$O(^3P) + CO(X^1\Sigma_g^+) + M \rightarrow CO_2(X^1\Sigma_g^+) + M + 532\ \text{kJ/mol} \quad (37)$$

The predominance of molecular oxygen in the atmosphere and the spin-conservation rules (which govern the probability of reaction between species in different spin states) favor instead the formation of ozone by Reaction 38, involving reaction of the electronic ground states of atomic oxygen [$O(^3P)$] and molecular oxygen [$O_2(^3\Sigma_g^-)$] in the presence of a chaperone molecule, M.

$$O(^3P) + O_2(^3\Sigma_g^-) + M \rightarrow O_3(^1A) + M \quad (38)$$

The relative efficiencies* of Reactions 37 and 38 at ambient temperatures in air are given by the ratio of the rates (R_{37}:R_{38}) of Reactions 37 and 38, respectively. Thus, where k_{37} and k_{38} are the rate coefficients of reaction for Reactions 37 and 38, respectively, [CO] = 10 ppm, [O_2] = 21%.

*The rate (R) of any elementary reaction is given by the product of the rate coefficient (k), at the temperature desired, and the concentrations of the reactants, in the same units (dimensions) as used in the rate coefficient, e.g., for the reaction A + B → C, $R_{-A} = R_{-B} = R_C = k[A][B]$, where R_{-A} and R_{-B} refer to the rates of removal of A and B, respectively, and R_C refers to the rate of formation of C.

$$\frac{R_{37}}{R_{38}} = \frac{k_{37}[CO]}{k_{38}[O_2]} = 4 \times 10^{-6}$$

Reaction 37 is fundamentally important at high temperatures in the combustion (flames and explosions) of carbonaceous materials and in the photochemistry of planetary atmospheres.

The kinetics and mechanisms of the reactions of oxygen atoms with carbon monoxide are reviewed and discussed in considerable detail elsewhere.[32,97,185,252] Although there have many studies of this reaction at room temperature and at the very high temperatures (2,300–3,600 K) in shock tubes, there still are large unresolved discrepancies in the reported kinetic measurements. The rate coefficients at room temperature range from 1.8×10^{-7} to $<5 \times 10^{-5}$ ppm^{-2} min^{-1}, and activation energies vary from -24 kcal/mol to $+4.5$ kcal/mol. Some studies have shown that the reaction rate is second-order dependent on reactant concentrations, but others have shown a third-order dependence, suggesting a termolecular process involving chaperone molecules (M). Additional studies are required to resolve these difficulties. The discrepancies may be due in part to: reaction conditions, wall effects, hydrogenous impurities generating hydroxyl radicals that react rapidly with carbon monoxide, or iron carbonyl impurities in the carbon monoxide used.

Simonaitis and Heicklen[406] have made a recent study of this reaction employing mercury photosensitization of nitrous oxide and competition for the oxygen atoms produced by carbon monoxide and 2-trifluoromethylpropene. They suggested the reaction proceeds via intermediate electronically excited states of carbon monoxide and proposed a mechanism to explain their observation that the reaction rate dependence was intermediate between second and third order in reactants.

$$O(^3P) + CO + M \rightleftarrows CO_2(^3B_2) + M \quad (39)$$

$$CO_2(^3B_2) \rightleftarrows CO_2(^1B_2) \quad (40)$$

$$CO_2(^1B_2) + M \rightleftarrows CO_2(^1\Sigma_g^+) + M \quad (41)$$

The reaction was found to be pressure dependent approaching second-order kinetics as the temperature was increased at any pressure. They derived expressions for limiting low- and high-pressure rate constants, k_0 and k_∞, respectively; $k_0 = 5.9 \times 10^9 \exp[-4{,}100/RT]M^{-2}\ s^{-1}$ (with nitrous oxides as the chaperone molecule M) and $k_\infty = 1.6 \times 10^7 \exp(-2{,}900/RT)M^{-1}\ s^{-1}$, where $k_0 = k_{39}$ and $k_\infty = k_{39}k_{40}/k_{-39}$ (k_{39} and k_{40} are the rate constants for the forward reactions in Equations 39 and

40 and k_{-39} is that for the reverse reaction in Equation 39). These values correspond to $k_0 = 5.7 \times 10^{-7}$ ppm^{-2} min^{-1} and $k_\infty = 2.9 \times 10^{-1}$ ppm^{-1} min^{-1} at ambient temperatures (25 C).

The bimolecular Reactions 42 and 43 also have been studied.

$$CO_2 + O(^3P) \rightarrow CO_2 + h\nu \quad (42)$$

$$CO + O(^3P) \rightarrow CO_2 \quad (43)$$

The rate constants evaluated for Reactions 42 and 43 are: $k_{42} = 8.3 \times 10^2 \exp(-2{,}590/RT) M^{-1} s^{-1} = 2.5 \times 10^{-5}$ ppm^{-1} min^{-1} at 25 C and $k_{43} = 1.8 \times 10^7 \exp(-2{,}530/RT) M^{-1} s^{-1} = 6.1 \times 10^{-1}$ ppm^{-1} min^{-1} at 25 C. Note the similarity between k_{43} and k_∞ for the previous set of processes (Reaction 39 to 41).

The reaction of carbon monoxide with electronically excited oxygen atoms, $O(^1D)$, is permitted by the spin-conservation rules, and Reactions 44 and 45 would therefore be expected to be considerably faster than the corresponding reactions with the ground-state atoms, $O(^3P)$.

$$CO(X^1\Sigma_g^+) + O(^1D) + M \rightarrow CO_2(X^1\Sigma_g^+) + M + 720 \text{ kJ/mol} \quad (44)$$

$$CO + O(^1D) \rightarrow CO_2 \quad (45)$$

Clerc and Barat,[86,87] using the far-ultraviolet flash photolysis of carbon dioxide, estimate that $k_{44} = 1$ ppm^{-2} min^{-1} and $k_{45} = (1 \pm 0.5) \times 10^4$ ppm^{-1} min^{-1} at about 300 K. However, the ambient concentrations of the excited species, $O(^1D)$, in the lower natural or polluted atmosphere are considerably less than those of the species $O(^3P)$, owing to a less efficient source of the former species (O_3 solar photolysis at $\lambda < 310$ nm) and to efficient removal of $O(^1D)$ by processes involving collisional de-excitation with molecules of oxygen, nitrogen, water, and argon in air and by chemical reaction with molecular oxygen and water vapor. Oxygen atoms, $O(^3P)$ and $O(^1D)$, are produced more efficiently at higher altitudes, and it is therefore probable that atomic oxygen can oxidize carbon monoxide at altitudes in the regions of the upper stratosphere and in the mesosphere where carbon dioxide itself can be photolyzed. In urban air and the lower stratosphere, the reaction with atomic oxygen is insignificant, as are the reactions of carbon monoxide with atomic hydrogen and organic free radicals because of the strong affinity of these transient species for atmospheric oxygen.

Hydroxyl radicals (OH) are generated by several recognized processes in the atmosphere, particularly in heavily polluted air resulting from automobile exhausts. These include: the solar photolysis of nitrous acid vapor (HONO) at $\lambda < 400$ nm; reaction of electronically excited oxygen atoms [$O(^1D)$] with water vapor by Reaction 46; abstraction of hydrogen from hydrocarbons by ground-state atomic oxygen [$O(^3P)$] as in Reaction 47; and indirectly by the solar photolysis of aldehydes ($\lambda < 350$ nm) followed by the reactions of the product species with atmospheric constituents and trace contaminants, oxygen, and nitric oxide, as shown by the sequence of Reactions 48 through 51.

$$O(^1D) + H_2O \rightarrow 2OH + 117 \text{ kJ/mol} \quad (46)$$

$$RH + O(^3P) \rightarrow R\cdot + \cdot OH + (59\text{–}84) \text{ kJ/mol} \quad (47)$$

$$RCHO \xrightarrow{h\nu} R\cdot + \cdot CHO \quad (48)$$

$$\cdot CHO \rightarrow \cdot H + CO - 100 \text{ kJ/mol} \quad (49)$$

$$H\cdot + O_2 + M \rightarrow HO_2\cdot + M + 197 \text{ kJ/mol} \quad (50)$$

$$HO_2\cdot + NO \rightarrow NO_2 + \cdot OH + 38 \text{ kJ/mol} \quad (51)$$

Photolysis of aldehyde (Reaction 49) is not a significant source of atmospheric carbon monoxide compared to its production by the internal combustion of gasoline–air mixtures in automobiles. In polluted atmospheres, the solar photolyses of aldehydes, which are partially oxidized hydrocarbons, by Reaction 48 and specifically that of formaldehyde by Reaction 52, are important sources of transient reactive intermediates such as hydrogen atoms (H), hydrocarbon radicals (R, RO, RO_2, RCO), and hydroperoxyl (HO_2) and hydroxyl (OH) radicals.

$$HCHO \xrightarrow{h\nu} \cdot H + \cdot CHO \quad (52)$$

The hydroxyl radical reacts rapidly with carbon monoxide at both ambient and subambient temperatures, forming carbon dioxide and atomic hydrogen.[124,174]

$$OH + CO \rightarrow CO_2 + \cdot H + 105 \text{ kJ/mol} \quad (53)$$

From an evaluation[32] of reported rate constants, the most reliable kinetic expression for Reaction 53 is $k_{53} = 5.6 \times 10^8 \exp(-1{,}080/RT)M^{-1}$ s^{-1}. Thus, large rate coefficients at ambient and subambient temperatures may be derived (i.e., $k_{53} = 2.2 \times 10^2$ ppm^{-1} min^{-1} at 300 K) because of the low activation energy (4.5 kJ/mol) of this reaction. This fast reaction between the hydroxyl radical and carbon monoxide is believed to be important in polluted atmospheres as well as in the upper atmosphere, where hydroxyl radicals are produced in quantity through Reaction 46. In polluted atmospheres, however, trace contaminants such as hydrocarbons (particularly olefins), sulfur oxides, and nitrogen oxides can compete successfully for available hydroxyl radicals and thus reduce significantly the probability for reaction with carbon monoxide.

It has been suggested[189,416,481] that the rapid conversion (half-life $\approx$1 hr) of nitric oxide to nitrogen dioxide in photochemical smog can be explained by the cyclic chain of Reactions 50, 51, and 53. In this sequence, hydroxyl radicals consumed by Reaction 53 are subsequently regenerated from the hydrogen atoms produced in Reaction 53 through the consecutive steps of Reactions 50 and 51. In this way repeated cycles regenerate and maintain ambient hydroxyl concentrations while simultaneously converting carbon monoxide and nitric oxide to carbon dioxide and nitrogen dioxide, respectively. The rate of conversion of nitric oxide to nitrogen dioxide would not be greatly affected by variations in ambient carbon monoxide concentrations if the carbon monoxide concentrations were always much higher than the nitric oxide concentration. This condition is usually met in urban pollution and elsewhere, when the carbon monoxide concentrations exceed those of nitric oxide by at least two orders of magnitude. The ambient carbon monoxide concentrations would be affected very slightly ($<1\%$) and immeasurably by involvement in these processes since the amount reacted would be less than the inherent error in standard monitoring systems, such as nondispersive infrared spectrophotometry.

Reaction 53 is a chemical sink for carbon monoxide in the stratosphere. Despite the low temperatures near the tropopause (213 K or -60 C), the high rate of this reaction explains the relatively rapid decrease in the mixing ratio of carbon monoxide above the tropopause (11 km at midlatitudes). Estimates of the atmospheric concentrations of hydroxyl radicals required to react with carbon monoxide by Reaction 53 have been made.[254,255]* Until recently, estimated concentrations for these chemical species in the range of 10^{-7}–10^{-8} ppm were below the limits of detection by the available methods. In 1976, it was reported that hydroxyl radical concentrations were measured in the upper stratosphere ranging from 4.5×10^6 cm^{-3} at 30 km to 2.8×10^7 cm^{-3} at 43 km, by

*G. J. Doyle (Stanford Research Institute, Menlo Park, California), unpublished data, 1968.

a molecular resonance fluorescence emission detection instrument.[7] Before mechanisms such as those cited above can be postulated with any degree of certainty, the concentrations of these and other reactive intermediates (atoms and free radicals), such as hydroperoxyl (HO_2) and nitrate (NO_3) have to be determined *in situ* in ambient urban atmospheres. Some progress is being made at the present time in developing techniques and instrumentation for this purpose,[26] but we are still a long way from being able to measure reproducibly and accurately all important reactants and products of either photochemical or other types of pollutants in urban environments. The formidable task of performing similar measurements at much lower concentrations to determine the background concentration of the relatively clean, natural atmosphere appears to be impracticable at this time.

Westenberg and deHaas[483] used electron spin resonance detection in a discharge flow system to investigate in the laboratory the competition between carbon monoxide and hydrogen atoms for hydroperoxyl radicals by Reactions 54 and 55, respectively.

$$CO + HO_2\cdot \rightarrow CO_2 + \cdot OH + 264 \text{ kJ/mol} \qquad (54)$$

$$\cdot H + HO_2\cdot \rightarrow 2\cdot OH + 159 \text{ kJ/mol} \qquad (55)$$

From analysis of their data, they claimed that Reaction 54 involving oxidation of carbon monoxide by the hydroperoxyl radical could be faster than oxidation by the hydroxyl radical, Reaction 53. Gorse and Volman,[167] on the other hand, used the static photolysis of hydrogen peroxide in the presence of carbon monoxide to estimate that the room temperature rate coefficient for Reaction 54 is at least 10 orders of magnitude less than predicted by Westenberg and deHaas. This large discrepancy must be resolved before it can be stated with confidence that a significant reaction exists between carbon monoxide and hydroxyl that can effectively compete with the conversion of nitric oxide to nitrogen dioxide,[482] as shown in Reaction 51.

There is some evidence that halogen atoms, from the photolysis of fluorine, chlorine, and bromine, may catalyze the oxidation of carbon monoxide[188,271] in oxygen at ambient temperatures (25 C). At this time, the extent to which these reactions would be of importance in the atmospere is uncertain. Participating halogenated species are produced and exist in the marine atmosphere and also in the upper atmosphere, where they are produced by the decomposition of halocarbon aerosol propellants and refrigerants as well as natural materials such as methyl chloride.

Chemical Modeling of Carbon Monoxide Reactions In the previous section, in which reactions of carbon monoxide were discussed, the processes described were isolated and studied by suitably selecting and controlling experimental conditions in order to determine the kinetic characteristics of the reaction system. In the atmosphere, however, the situation is complicated by the enormous variety of competing and consecutive processes and interactions that occur. As a result, simple assumptions cannot provide a reliable picture of the transformations occurring, nor can the influence by reliably assessed of the effects of variations in one contaminant on the rates of conversion and levels of concentration achieved by the other contaminants. In order to assess the influence of varying one or more parameters in the atmospheric reaction system, it has been necessary to develop theoretical numerical models that incorporate the most recent kinetic information on all the significant chemical reactions. Mathematical models have come to prominence with the advent of high-speed and large-capacity computers, which make handling masses of data efficient and manageable. The initial efforts to develop reliable and realistic models have been in the analysis of smog chamber data obtained by simulating synthetic photochemical pollution reactions at near ambient concentrations of reactants. As these models are tested and refined, it should be possible eventually to achieve a reliable predictive model combining chemistry, physics, and meteorology that can be used to assess the design and impact of various pollution control strategies. Present models, however, are generally capable of describing reactions only in simple mixtures under carefully controlled conditions and are of limited atmospheric application due to the sensitivity to chemical input, rather than because of the model design. These models make their greatest contributions by the identification of the limitations of smog chamber experiments and by indicating the lack of knowledge about many of the elementary chemical processes involved.

Theoretical considerations suggest that carbon monoxide as well as hydrocarbons can be involved in the conversion of nitric oxide to nitrogen dioxide in polluted atmospheres, where Reactions 50, 51, and 53 can be represented by Reaction 56:

$$CO + NO + O_2 \rightarrow CO_2 + NO_2 \quad (56)$$

Modeling computations indicate that carbon monoxide may play a role in ozone production.[480] Ozone concentrations were determined to be 50% higher in the presence of 100 ppm carbon monoxide than calculated for the same hydrocarbon–nitrogen oxide–air mixture with no carbon monoxide present. This suggests that substantial reductions in carbon monoxide emissions could reduce ozone production in photochemical smog, but this has been questioned,[157] at least for carbon monoxide concentrations <35 ppm. Also, because these simulations were performed using a single hydrocarbon reactant, isobutylene, in a controlled environment, this conclusion cannot be extrapolated with certainty to the complex blend of organic compounds found in automobile exhausts and the conditions existing in urban atmospheres.

Calvert *et al.*[64] have developed a model to simulate a simple analogue of the sunlight-irradiated auto-exhaust-polluted atmosphere. They concluded that although Reactions 50, 51, and 53 constitute a major regeneration step for hydroperoxyl radicals in the chain oxidation of nitrogen dioxide, it is by no means the main source of hydroxyl-to-hydroperoxyl conversion in their model. They suggest that the H atom abstraction reaction of alkoxy radicals with oxygen provides a more efficient source. In a computer simulation study of the effect of carbon monoxide on the chemistry of photochemical smog, the same authors[65] conclude that the presence of small levels of carbon monoxide in a nitric-oxide-containing atmosphere can enhance the photooxidation of nitric oxide to nitrogen dioxide, ultimately forming significant amounts of ozone. Nevertheless, there are still many discrepancies between the effect predicted by chemical mathematical simulation models and the results from experimental analyses of reacting mixtures in smog chambers. These discrepancies will be resolved as our knowledge about these elementary chemical processes increases, as the analytical techniques improve, and as the models are subjected to more rigorous testing and refinement.

In the last few years several photochemical models of the natural troposphere have been developed.[112,210,245,254,255,256,466] Using plausible reaction schemes, Levy[254,255] has demonstrated that radical reactions are important in this region of the atmosphere. He has shown specifically that hydroxyl radicals achieve significant concentrations in the sunlit atmosphere and that their subsequent reactions with trace gaseous constituents, including carbon monoxide, could be important. The hydroxyl radical is essentially unreactive with the major atmospheric components (nitrogen, oxygen, argon, and carbon dioxide); this accounts for its apparent preference for reactions with minor constituents. McConnell *et al.*,[293] using a model similar to that of Levy, estimated that at noontime the hydroxyl concentration in the lower regions of the troposphere was 8×10^{-8} ppm, with a daily average of about 2×10^{-8} ppm. The latter value decreases with increasing altitude to about one-third this concentration in the vicinity of the tropopause.

According to Levy,[254,255] the major source of hydroxyl radicals in the lower troposphere is the reaction of electronically excited oxygen [$O(^1D)$] with water vapor, shown in Reaction 46. The oxygen species [$O(^1D)$] is produced by the efficient solar photolysis of ozone at the long wavelength at the end of the Hartley absorption band (λ = 290–350 nm). Although most of the excited oxygen [$O(^1D)$] atoms are converted to the ground state [$O(^3P)$] by collisions with atmospheric nitrogen and oxygen molecules, a few percent still are available to react with water vapor. The concentrations of hydroxyl radicals are maintained by the processes discussed previously (Reactions 46, 47, and 50). An important consequence of these observations is that hydroxyl radicals react rapidly with methane (CH_4),[113,210,255,293] as well as with carbon monoxide, and at a sufficiently high rate to provide the most significant natural source of carbon monoxide identified thus far. The primary step in the initiation of atmospheric oxidation of methane involves hydrogen abstraction by the hydroxyl radical by Reaction 57.

$$CH_4 + OH \rightarrow CH_3 + H_2O + 71 \text{ kJ/mol} \tag{57}$$

The rate coefficient determined[174] for Reaction 57 is k_{57} = 12 ppm^{-1} min^{-1} at 25 C. In conjunction with a methane concentration of 1.4 ppm and the average hydroxyl concentration given previously, the upper limit for carbon monoxide production is found to be $R_{CO} = 3.4 \times 10^{-7}$ ppm min^{-1}, assuming that all CH_4 molecules reacted are converted to carbon monoxide. To produce carbon monoxide, Reaction 57 is followed by reaction of the methyl radicals (CH_3) with molecular oxygen to produce formaldehyde (CH_2O) by Reactions 58 and 59a and formyl radicals (CHO) by Reaction 59b. The solar photolysis of formaldehyde produced by Reaction 59a is an important natural source of carbon monoxide by Reaction 60a and of hydrogen atoms and formyl radicals by Reaction 60b.

$$CH_3 + O_2 \xrightarrow{(M)} CH_3O_2\cdot \tag{58}$$

$$CH_3O_2\cdot \rightarrow CH_2O + \cdot OH \tag{59a}$$

$$CH_3O_2\cdot \rightarrow \cdot CHO + H_2O \tag{59b}$$

$$CH_2O \xrightarrow{h\nu} H_2 + CO \tag{60a}$$

$$CH_2O \xrightarrow{h\nu} \cdot H + \cdot CHO \tag{60b}$$

Formaldehyde is a reactive and photochemically unstable intermediate that may efficiently produce carbon monoxide in the absence of oxygen. Several authors[113,255,293,473] have estimated that the oxidation of biologically produced methane leads to a global production rate of carbon monoxide at least 10 times greater than that obtained from anthropogenic sources. Recently, however, Warneck[466] has proposed that Levy's model should be modified to take into account the possible scavenging of hydroperoxyl radicals (an intermediate in the hydroxyl recycling process) by atmospheric aerosols.

Numerous photochemical models also have been used to describe stratospheric processes. These models use sets of sequential reactions similar to those used to describe photochemical smog but with appropriate modifications for different temperatures, pressures, concentrations, and solar spectral intensity distributions. In the stratosphere the dominant scavenging process for removal of carbon monoxide is the reaction with the hydroxyl radical.

3

Sources, Occurrence, and Fate of Atmospheric Carbon Monoxide

Carbon monoxide is released into the atmosphere from both natural and anthropogenic sources. Recent theoretical estimates[113,255,293,473] suggest that on a global basis natural sources contribute at least 10 times more carbon monoxide than man-made sources to the total atmospheric burden. The most important natural source that has been suggested[255,293] is that resulting from the oxidation of atmospheric methane, with lesser contributions from forest fires, terpene oxidation, and the oceans. The incomplete combustion of fossil fuels is the principal anthropogenic source of carbon monoxide. At the present time, carbonaceous fuels are widely used, primarily in transportation and to a lesser extent in space heating and industrial processing. In contrast to natural sources, which are globally widespread, anthropogenic sources are mainly located in urban and metropolitan areas and concentrated in the Northern Hemisphere. The influence of naturally produced carbon monoxide on the carbon monoxide concentration in urban air is believed[210] to be negligible, but recent discoveries indicate that it could be important in determining the "background" concentrations of carbon monoxide and the mean residence time (atmospheric lifetime) in the terrestrial atmosphere.

Anthropogenic carbon monoxide production is directly related to man's technological growth and productivity, as well as to his economic and social well-being. It is well within our present technologic capability to reduce the emission of this pollutant, and maintain it at a low level without economic hardship, despite its being an unwelcome by-product accompanying national economic growth along with increasing fossil fuel energy requirements. There appears to be no simple panacea, however, for controlling pollutant emissions, as shown by the new pollution problems that appear when specific controls are implemented. Much more attention and research have to be devoted in the future to the indirect as well as the obvious consequences of applying control techniques to both stationary and mobile sources.

It is estimated that the total anthropogenic emission of carbon monoxide exceeds that of all other man-made pollutants combined. This fact, coupled with the purported excessive natural emission of carbon monoxide and its long lifetime (residence time) owing to its stability and apparent lack of chemical and biologic reactivity, supports the conclusion that carbon monoxide is the most abundant and most commonly occurring of all the atmospheric pollutants* and minor constituents.

SOURCES OF CARBON MONOXIDE

Contrary to what has been believed,[331,450] it is now considered possible that atmospheric carbon monoxide could be principally of natural origin.[255,293] This gas was first discovered as a trace constituent of the terrestrial atmosphere by Migeotte[311] in 1949. In a study of the solar spectrum, he attributed specific absorption lines to ambient carbon monoxide originating at a wavelength of about 4.7 μm in the infrared region. Similar observations[38,218,272,312] were made during the next decade in the United States, Canada, and Europe that demonstrated that carbon monoxide concentration levels average about 0.1 mg/m^3 (0.1 ppm) in clean air. From these early investigations, it was concluded by Junge[218] in 1963 that background atmospheric carbon monoxide concentrations were highly variable and erratic in nature but generally ranged in concentration from 0.01 to 0.2 ppm.

*Although carbon dioxide (CO_2) is produced and released to the atmosphere in much greater quantities than carbon monoxide, it is not usually classified or defined as an air pollutant. Little can be done at this time to control carbon dioxide emissions despite a regularly observed and significant (0.7–1.0 ppm/yr) annual increase in the already large concentrations of global ambient carbon dioxide (330 ppm) or the associated global climatic implications should the annual-increase trend continue. This gas is a natural end product of all combustion processes involving carbonaceous materials. Thus, not until it is economically and practically feasible to replace conventional fossil-fuel-burning energy sources with noncombustive systems such as nuclear, solar, or geothermal power and provide noncarbonaceous fuels for transportation will it be possible to markedly reduce carbon dioxide emissions.

In 1968, Robinson and Robbins[378] estimated that about 2.8×10^8 metric tons (2.8×10^{11} kg) of carbon monoxide were discharged into the atmosphere worldwide from anthropogenic sources during the year of 1966. Somewhat more than half of the total, 1.5×10^8 metric tons (1.5×10^{11} kg), was estimated to be generated in the continental United States. Transportation is by far the largest single man-made source of carbon monoxide in this country, accounting for more than two-thirds of the emissions from all anthropogenic sources in the 1960's. It has been estimated that the total U.S. emissions of carbon monoxide showed almost a twofold increase during the period 1940 through 1968. This dramatic increase was due almost exclusively to the increasing use of the motor vehicle during that period. After 1968 there was an initial decline of about 10% in the carbon monoxide emission followed by a leveling off in mobile source emission inventories owing to the required installation of emission control devices in new vehicles. In 1974, there were about 125 million vehicles registered in the United States,[449] which produced annually about 10^8 metric tons (10^{11} kg) of carbon monoxide.

The magnitude of the problem in controliing internal combustion engine emissions may be seen from the fact that approximately 3 lb (1.4 kg) of carbon monoxide, along with other pollutants, is produced in the combustion of 1 U.S. gal (3.8 liter) of gasoline. Assuming an average density for gasoline (octane, C_8H_{18}) of 0.7 g/cm^3, then, on the basis of mass, approximately 25% of the carbon in gasoline is converted to carbon monoxide in the internal combustion engine. This corresponds to an oxidation efficiency of 92%, assuming carbon dioxide and water are the only other products. If the average automobile (without emission control equipment) travels 15 miles (24 km) per gallon of fuel consumed, then about 0.2 lb (91 g) of carbon monoxide per mile (52 g/km) would be released into the surrounding atmosphere, which was the situation prior to 1968. This average emission is two and one-half times greater than the 1970–71 automobile emission standard set for carbon monoxide (34 g/mile, 20 g/km), but more significantly it is 25 times larger than the proposed standard for 1975–76 model automobiles (3.4 g/mile, 2.0 g/km), as prescribed by the 1970 Clean Air Amendments. It should be pointed out here that the required installation of catalytic converters in 1975 model automobile exhaust systems in conjunction with the use of unleaded gasoline has greatly reduced carbon monoxide emissions; but the reduced emissions of criteria pollutants were achieved at the expense of generating potentially harmful sulfuric acid aerosol. Apparently there is no simple solution to the problems of automobile emission control.

Technological Sources

Until recently, anthropogenic sources producing large quantities of carbon monoxide were considered to be primarily responsible for the observed[331,450] global concentrations. Although carbon monoxide produced in urban areas by man's activities far surpasses natural contributions, it now appears that anthropogenic emissions only approximate natural contributions in the Northern Hemisphere, and on a global scale man's activities contribute only about 10% to the total worldwide production. Estimates[211,378] of global tonnages of carbon monoxide produced by anthropogenic sources show a marked increase in annual emissions, which correlates with the worldwide increase in the consumption of fossil fuels. A comparison of a recent estimate[211] of global carbon monoxide emissions from man-made sources with an earlier estimate[378] shows that annual worldwide anthropogenic carbon monoxide emissions have increased by 28% during the period 1966–70. Jaffe[211] has estimated that in 1970 the global carbon monoxide emissions from combustion sources were about 359 million metric tons (4.0×10^8 short tons or 3.6×10^{11} kg). This estimate is based on an analysis of worldwide fossil fuel usage, agricultural practice, mining activities, and waste disposal in conjunction with recently revised (1972) carbon monoxide emission factors for various sources. Table 3-1 gives the estimated worldwide contributions from major sources plus the corresponding fuel consumption figures. These data show that, as in previous estimates, the internal combustion engine in motor vehicles still represents the largest single source of carbon monoxide; it provides 55% of all anthropogenic emissions. The second most important anthropogenic sources are industrial processes and certain miscellaneous activities in the stationary source category, which together contribute almost 25% of the total. In 1970 the U.S. contribution was 37% of total global production from all anthropogenic sources.

During the period 1940–68, there was a dramatic increase in carbon monoxide emissions in the United States due almost exclusively to the increased use of automobiles. Since 1968, automobile emissions have leveled off through 1970 and declined from 1970 through 1975 because of the installation of emission control devices. Trends in the emissions from major sources for the period 1940–75 are shown in Table 3-2. During the 35-yr period covered in Table 3-2, transportation in the United States was clearly the dominant contributor to the total carbon monoxide emissions, as shown by the increasing percentage from this source, from 41% in 1940 to 77% in 1975. The carbon monoxide emissions from industrial sources also were greatly reduced during the period 1940

TABLE 3-1 Estimated Global Anthropogenic Carbon Monoxide Sources for 1970[a]

Source	World Fuel Consumption, 10^6 metric tons/yr (10^9 kg/yr)	World Carbon Monoxide Emission, 10^6 metric tons/yr (10^9 kg/yr)
Mobile		
Motor vehicles	439	197
gasoline and diesel		2
Aircraft (aviation gasoline, jet fuel)	84	5
Watercraft		18
Railroads		2
Other (nonhighway) motor vehicles (construction equipment, farm tractors, utility engines, etc.)		26
Stationary		
Coal and lignite	2,983	4
Residual fuel oil	682	<1
Kerosene	69	<1
Distillate fuel oil	411	<1
Liquefied petroleum gas	34	<1
Industrial processes (petroleum refineries, steel mills, etc.)		41
Solid-waste disposal (urban and industrial)	1,130	23
Miscellaneous (agricultural burning, coal bank refuse, structural fires)		41
Total anthropogenic carbon monoxide		359

[a]Reprinted with permission from Jaffe[211] (L. S. Jaffe, Carbon monoxide in the biosphere: Sources, distributions, and concentrations, J. Geophys. Res. 78:5293–5305, 1973. Copyrighted by American Geophysical Union).

through 1970, although there was a rapid industrial expansion during this period. Agricultural burning and industrial process losses, representing 9.3% and 7.6%, respectively, of the total carbon monoxide emissions, are two important sources that have remained relatively constant until 1970. A considerable reduction (77%) in emissions from stationary sources occurred during the period from 1940 through 1975, probably

TABLE 3-2 Nationwide Estimates of Carbon Monoxide Emissions, 1940–75[72,455,456]

	Emissions, 10^6 metric tons[a]/yr (10^9 kg/yr)							
Source Category	1940	1950	1960	1968	1969	1970	1972	1975[b]
Transportation	31.7	50.2	75.7	102.5	101.6	100.6	70.4	66.5
Industrial process losses	13.1	17.1	16.1	7.7	10.9	10.3	15.8	13.3
Agricultural burning	8.3	9.4	11.2	12.6	12.5	12.5	1.5	0.8
Fuel combustion in stationary sources	5.6	5.1	2.4	1.8	1.6	0.7	1.1	1.3
Solid-waste disposal	1.6	2.4	4.6	7.3	7.2	6.6	4.5	3.4
Miscellaneous	17.2	9.1	5.8	4.9	5.7	4.1	4.2	1.7
Total	77.5	93.3	115.8	136.8	139.5	134.8	97.5	87.0

[a]To convert to U.S. short tons, multiply by 1.1 (1 short ton = 2×10^3 lb).
[b]Annual emission as of March 12, 1975.

owing both to a changeover from solid fossil fuels to liquid and gaseous fuels and to increased furnace efficiency. All of the sources listed can be controlled with the exception of those in the miscellaneous category. The latter include such essentially uncontrollable sources as forest fires, structural fires, and burning in coal refuse banks.

Table 3-3 gives a more detailed breakdown of emission sources with estimates of the amounts of carbon monoxide emitted[72,211] during calendar year 1975. These data show that in 1975 the carbon monoxide emissions in the United States from technological sources (all categories except forest fires) was more than 85×10^6 metric tons (85×10^9 kg, 93×10^6 short tons). The major source of carbon monoxide in the United States in 1975 was the combustion of fossil fuels in vehicles producing about 67 million metric tons (67×10^9 kg, 74×10^6 short tons) annually, with gasoline-burning internal combustion engines accounting for 65% of the total.

At the present time, the remaining sources in order of decreasing contribution are industrial processes, solid-waste combustion, miscellaneous fires, and agricultural burning. Fuel combustion in stationary sources for space heating and power generation provides less than 2% of the total carbon monoxide emissions and, thus, is of minor importance. Carbon monoxide is also locally produced in high concentration by burning cigarettes, explosions, and the firing of weapons, but these point sources are insignificant in terms of total annual production.

Natural Sources

Prior to the 1970's, the known natural sources of carbon monoxide were considered to be of minor importance in comparison with technological sources, and the principal natural source was thought to be forest fires re-

TABLE 3-3 Detailed Summary of 1975 Carbon Monoxide Emission Estimates in the United States[455]

Source Category	Estimated Carbon Monoxide Emissions, 10^3 metric tons[a]/yr (10^6 kg/yr)
Fuel combustion in stationary sources	
Steam-electric	255
Industrial	468
Commercial and institutional	71
Residential	457
Total fuel	1,251
Transportation	
Gasoline vehicles	57,226
Diesel vehicles	553
Total road vehicles	57,779
Railroads	270
Vessels	970
Aircraft	778
Other nonhighway use	6,732
Total transportation	66,529
Solid-waste disposal	
Municipal incineration	206
On-site incineration	1,893
Open burning	1,307
Total solid waste	3,406
Industrial process losses	13,278
Agricultural burning	773
Total controllable	85,237
Miscellaneous	
Forest fires	1,640
Structural fires	30
Total miscellaneous	1,670
Total all categories	86,907

[a]To obtain U.S. short tons, multiply by 1.1 (1 short ton = 2×10^3 lb).

sulting from natural causes, such as lightning.[378] Other natural emission sources identified were volcanic activity, natural gases from marshes and coal mines,[140] and electrical storms.[485] Secondary sources of natural origin included photochemical degradation of naturally occurring organic compounds, such as aldehydes, which are also involved in photochemical smog formation,[4,5,252] and the solar photodissociation of carbon dioxide, which becomes feasible in the upper atmosphere at altitudes above 70 km.[31]

In the last few years, several potentially large natural sources of geophysical or biological origin have been identified. Estimates of the magnitude of these sources range from 3 to 25 times that of anthropogenic sources, depending upon the data base selected for comparison.

In 1972, Stevens *et al.*[417] reported a comprehensive study made of the isotopic composition of atmospheric carbon monoxide at numerous locations at different times of year, which included a comparison of the carbon monoxide originating from automobile exhausts with that from selected natural sources. They found five major isotopic types of carbon monoxide, two containing light oxygen (^{16}O enriched) and three containing heavy oxygen (^{18}O enriched). The light-oxygen varieties, which were present throughout the year, were a predominant fraction of atmospheric carbon monoxide, with constant concentrations of 0.10–0.15 ppm. The heavy-oxygen species were minor constituents whose production was apparently seasonal. On the basis that the light-oxygen species originate in the atmosphere rather than the biosphere, they estimated an atmospheric carbon monoxide production rate in the Northern Hemisphere of more than 3×10^9 metric tons (3×10^{12} kg). From the estimate of Robinson and Moser[376] that 95% of the global anthropogenic carbon monoxide emissions originate in the Northern Hemisphere and the total global man-made carbon monoxide emitted in 1970 given in Table 3-1, the man-made emissions in the Northern Hemisphere in 1970 are calculated to be about 3.4×10^8 metric tons, which is 10 times lower than the estimate from the isotopic studies.

Methane[293,473,493] and formaldehyde[66] have been suggested also as natural atmospheric sources of carbon monoxide. McConnell *et al.*[293] have estimated that atmospheric oxidation of biologically produced methane can provide a global source of about 2,500 million metric tons (2.5×10^{12} kg) annually. This is 10 times greater than the previously reported[378] worldwide anthropogenic emission rates and 7 times larger than the estimates for 1970 given in Table 3-1. In 1972, Weinstock and Niki[473] derived a carbon monoxide production rate via methane oxidation based on calculations of hydroxyl radical concentrations in tropospheric air. They concluded that this mechanism could provide a source strength about 25 times greater than man-made sources.

In some regions, the oceans appear to be a significant source of carbon monoxide. Independent studies[398,431] of Atlantic Ocean surface waters have shown that the surface layers are supersaturated with carbon monoxide ranging from about 10 to 40 times the equilibrium water-air ratio. Similar conclusions have been drawn from subsequent marine air-water interface studies in the Atlantic[248,430,489] and South Pacific oceans.[428] In 1971, Junge *et al.*[219] calculated that the oceans may contribute the equivalent of about 0.3% of the total anthropogenic carbon monoxide production. More recently, Linnenbom *et al.*[269] have revised their previous estimates of the oceanic carbon monoxide production to a Northern Hemisphere flux of 9×10^7 metric tons per year, or about 25% of the man-made output in 1970. In 1974, Liss and Slater[270] developed a mathematical model describing the flux of various gases across the air-sea interface. Taking a mean atmospheric carbon monoxide concentration of 0.13 ppm[248] and a surface water concentration of 6×10^{-8} cm^3 CO/cm^3 H_2O in their model, they estimate a total oceanic flux of 4.3×10^7 metric tons (4.3×10^{10} kg) of carbon monoxide per year. All these recent estimates suggest that the oceans are a source rather than, as previously thought, a sink for atmospheric carbon monoxide. Biological organisms including marine algae, siphonophores, and microorganisms apparently are responsible for the large quantities of carbon monoxide in the surface layers of the oceans.

In 1960, Went[479] proposed that atmospheric photochemical reactions involving naturally produced terpene hydrocarbons could be an important source of atmospheric trace constituents. Revised estimates of natural terpene production[478] were used by Robinson and Moser[376] to calculate that approximately 54×10^6 metric tons (5.4×10^{10} kg) of carbon monoxide are produced annually by atmospheric photochemical oxidation. An identical figure (5.4×10^{10} kg/yr) for carbon monoxide generation from the degradation of chlorophyll was arrived at by Crespi *et al.*[111] This source strength was used by these investigators together with an estimate of production by the bilin biosynthesis, which takes place in blue-green algae, to show that plants could produce a total of 90 million metric tons (9×10^{10} kg) of carbon monoxide per year on a global scale.

Charged particle deposition mechanisms and atmospheric electrical discharge phenomena, including lightning in the troposphere, have been investigated[172] as potential sources of atmospheric carbon monoxide. However, these sources appear to be small compared with anthropogenic sources.

Total global emissions of carbon monoxide can be estimated from anthropogenic and natural sources given in Table 3-1 and from the data presented above. Annual production rates for the major sources are given in Table 3-4. These estimates show that approximately 3.2 billion metric tons (3.2×10^{12} kg) of carbon monoxide can be released annually into the atmosphere by all processes.

OCCURRENCE OF CARBON MONOXIDE

Community Atmospheres

Carbon monoxide concentrations in metropolitan areas vary considerably both temporally and spatially. Analysis of aerometric data collected at continuous air monitoring program (CAMP) stations in selected cities has revealed distinct temporal patterns in ambient carbon monoxide levels. Diurnal, weekly, and seasonal trends have been observed which correlate with traffic volume, vehicle speed, and meteorological conditions.[294]

There is a great variation in the carbon monoxide concentrations in urban metropolitan areas, ranging from 1 to more than 140 ppm, with the higher concentrations being observed as brief peaks in dense traffic. As a result, persons in moving vehicles subject to heavy traffic conditions can be exposed to carbon monoxide concentrations greater than 50 ppm for sustained periods. Larsen and Burke[249] have developed a mathematical model to statistically analyze and review the extensive aerometric data collected at numerous sampling locations in many large cities. They estimated that maximum annual 8-hr average carbon monoxide concentrations were approximately 115 ppm in vehicles in heavy traffic downtown, 75 ppm in vehicles operating on expressways or arterial routes, 40 ppm in central commercial and mixed industrial areas, and

TABLE 3-4 Estimated Carbon Monoxide Production Rates from Natural and Anthropogenic Sources, 1970

Source	CO Emission Rate, 10^6 metric tons/yr (10^9 kg/yr)	Reference
Anthropogenic	359	211
Methane oxidation	2,500	293
Forest fires	10	378
Terpene oxidation	54	376
Plant synthesis and degradation	90	211
Oceans	220	269
Total, all carbon monoxide sources	3,233	

about 23 ppm in residential areas. These estimates indicate that carbon monoxide can achieve levels in heavy traffic on city streets almost three times that found in central urban areas and five times that found in residential areas. Mathematical urban diffusion models for carbon monoxide have been developed at Stanford Research Institute[339] and elsewhere in order to describe better the observed spatial and temporal variations of this pollutant.

In special situations, such as in underground garages, tunnels, and loading platforms, carbon monoxide levels have been found to exceed 100 ppm for extended periods.[107] To prevent excessive exposure to carbon monoxide, alarm systems have been installed in newly constructed tunnels that automatically activate auxiliary ventilating units when predetermined undesirable carbon monoxide levels are reached.

Background Levels and Distribution of Carbon Monoxide

Ambient carbon monoxide concentrations measured in relatively clean air (remote from strong sources) are quite low but variable. Some of the early measurements of the infrared solar spectrum absorption by atmospheric carbon monoxide at locations in Canada,[272] Switzerland,[44,311,312] and the United States[272,402,403] showed a range of concentrations from 0.03 to 0.22 ppm carbon monoxide and an average concentration[402] of 0.11 ppm.

Junge[218,219] reported that background concentrations of carbon monoxide range from 0.01 to 0.2 ppm. The most extensive measurements of carbon monoxide background levels at various locations have been made by Robinson, Robbins, and their colleagues. They have found concentrations ranging as low as 0.025 ppm and up to 0.8 ppm in North Pacific marine air;[374] 0.04 to 0.8 ppm carbon monoxide in nonurban air over California;[374] 0.06 to 0.26 with an average of 0.09 ppm at Point Barrow, Alaska;[71] 0.05 to 0.7 with an average of 0.11 ppm at Inge Lehmann Station in Greenland;[377] and an average of 0.06 ppm in the Southern Pacific.[377] They have concluded from these investigations that the observed variability of carbon monoxide in unpolluted areas is a characteristic of the air mass in transit and reflects the prior history of the air mass. Background levels as high as 1.0 ppm are observed when the air mass has recently traversed densely populated areas. On the basis of data collected on five cruises in the Pacific Ocean,[377] they measured and determined the average background concentration distributed latitudinally over the Pacific. The highest concentrations, about 0.2 ppm, were found between 30° and 50° N. These were associated with the high population density at midlatitudes in the Northern Hemisphere. The Northern Hemisphere midlatitude value decreases to about 0.07 ppm in the Arctic and to 0.09 ppm at the equator. In going south from the equator, the average concentrations fall to a minimum of about 0.04 ppm at 50° S and then rise to about 0.08 ppm in the Antarctic. Based on this work, they suggest average carbon monoxide concentrations of 0.14 ppm in the Northern Hemisphere, 0.06 ppm in the Southern Hemisphere, and a global average of 0.1 ppm.

Seiler and Junge[398] have measured similar average values over the North (0.18 ppm) and South (0.05 ppm) Atlantic Ocean. Generally higher (>30%) background concentrations have been found over the Atlantic Ocean than over the Pacific Ocean by several investigators. In general, because of its remoteness from major polluting sources, the Pacific Ocean may be considered to be more closely representative of "background" clean air.

The altitude and vertical distributions of atmospheric carbon monoxide have been reported by Seiler and Junge[398] and Robinson and Robbins.[377] They have found consistent carbon monoxide concentrations averaging about 0.13 ppm at 10 km altitude in both the Northern and Southern Hemispheres, in contrast to the large differences observed near the surface. During polar flights Robinson and Robbins observed[377] upper tropospheric concentrations averaging 0.10 ppm, whereas the stratosphere concentrations were markedly lower, falling in the range of 0.03 to 0.05 ppm carbon monoxide. Subsequently, Seiler and Warneck[400] found a decrease in concentrations from ~0.15 ppm below the tropopause to ~0.05 ppm above this region of discontinuity. A 1972 investigation[160] of the infrared solar spectrum made with a balloonborne spectrometer has shown a gradual decrease in carbon monoxide concentration with increasing altitude, from ~0.08 ppm at 4 km to 0.04 ppm at 15 km. These observations coupled with those made near the earth's surface indicate that important contributions to surface carbon monoxide result from the concentration of Northern Hemisphere industrialization at midlatitudes.

Residence times (τ = atmospheric mean life) for carbon monoxide in the atmosphere have been estimated by many investigators using the available data. An approximate estimate of this temporal turnover parameter can be made using the global average carbon monoxide concentration to calculate the total amount in the atmosphere (M_{CO}) and the total rate of production from natural and anthropogenic sources (R_{CO}). Thus, $\tau_{CO} = M_{CO}/R_{CO}$. Using the global annual production rate of 3.2×10^9 metric tons (from Table 3-4) and an average global concentration of 0.1 ppm, corresponding to 580×10^6 metric tons of carbon monoxide in the atmosphere, then $\tau = 5.8 \times 10^8 \text{ ton}/3.2 \times 10^9 \text{ ton/yr} = 0.18 \text{ yr}$.

These rough estimates are sensitive to the variability in and interpretation of atmospheric concentration measurements and to the current knowledge of source concentrations. They are, however, a useful guide in judging the effectiveness of natural processes in cleansing the atmosphere of this man-made pollutant and natural constituent. The early estimates of residence times for atmospheric carbon monoxide ranged from 2.7 to 5 yr.[218,374,378,429] However, in 1969, Weinstock[471] proposed that a residence time could be derived from radiocarbon data because "hot" carbon-14 (^{14}C) nuclei produced from the nitrogen-14 (^{14}N) (n, p) reaction with cosmic ray neutrons are fixed primarily as carbon monoxide-^{14}C (^{14}CO) prior to conversion into carbon dioxide-^{14}C ($^{14}CO_2$). From an analysis of available data on carbon monoxide-^{14}C (^{14}CO) concentrations in the atmosphere and estimates of its rate of formation, Weinstock derived a residence time of 0.1 yr, which was much shorter than the earlier estimates. Such a short atmospheric mean lifetime suggested that an efficient mechanism for carbon monoxide removal must be operative, and Weinstock proposed reaction with the hydroxyl radical.[471] In 1971, Levy[254] derived plausible theoretical estimates of tropospheric hydroxyl radical concentrations and their reaction with atmospheric carbon monoxide and calculated a corresponding residence time of 0.2 yr. The reaction of tropospheric hydroxyl radicals with atmospheric methane as the principal natural source of carbon monoxide was used by McConnell *et al.*[293] to estimate a residence time of 0.3 yr. Subsequently, Weinstock *et al.*[472,473] confirmed their earlier estimate of 0.1 yr from an analysis of both radioactive and stable carbon monoxide in the troposphere. There is sufficient evidence now to suggest that the residence time of carbon monoxide in the troposphere is about 0.2 ± 0.1 yr.

FATE OF ATMOSPHERIC CARBON MONOXIDE

On a global basis, the average concentration of carbon monoxide in the atmosphere is about 0.1 ppm (0.12 mg/m^3), and it does not appear to have been increasing substantially in recent time.[374] If efficient natural removal processes (sinks) were not operative to transform or scavenge this gaseous atmospheric pollutant, then based on the 1970 estimated total inputs of carbon monoxide (Table 3-4), the global average concentration would be increasing by about 0.5 ppm annually. Similarly, using current anthropogenic production estimates alone, atmospheric carbon monoxide would be expected to increase at the rate of 0.06 ppm/yr. As this does not appear to be the case, efficient sinks for carbon monoxide must be present in order to maintain the present atmospheric concentrations. The most important sink for atmospheric carbon monoxide identified at the present time appears to be the chemical reaction with ambient hydroxyl free radicals forming carbon dioxide. In addition, however, several possible natural sinks have been identified in the upper atmosphere, the biosphere, and the chemosphere. These will be discussed in more detail below.

Upper Atmosphere Sink

Carbon monoxide, produced at the earth's surface or in the lower troposphere, could conceivably migrate vertically to the upper regions of the atmosphere by atmospheric transport, turbulent mixing, and diffusion. A rapid decrease in carbon monoxide mixing ratio* above the polar tropopause was first observed by Seiler and Junge[399] in 1969. They attributed the reduction in carbon monoxide across the tropopause to the reaction between carbon monoxide and ambient hydroxyl radicals in the lower stratosphere. In 1970, theoretical estimates of the hydroxyl radical concentrations necessary to oxidize the flux of carbon monoxide from the troposphere into the stratosphere were reported to be 10^{-7} ppm.[194,358] Subsequently, flights were made in the winter of 1970–71 to measure the gradient of the carbon monoxide mixing ratio above the tropopause.[400] This sink appears to be controlled by the vertical eddy-type diffusion of carbon monoxide, and accounted for about 13 ± 2% of the total annual carbon monoxide based on 1968 estimates. This sink does not, therefore, appear to be adequate to compensate for carbon monoxide production at the present time.

Soil as a Sink

Certain microorganisms in the soil, as well as some terrestrial plant species, appear to be the major biological sinks for carbon monoxide. Soils both produce and absorb carbon monoxide simultaneously, but the net result is that they act as a sink. Ingersoll *et al.*[206] measured soil carbon monoxide uptake rates ranging from 21.1 to 319.4×10^{-11} g $CO/cm^2/s$ at 100 ppm carbon monoxide. This study, which was based on investigations in the field at 59 locations in North America,[205] showed that desert soils took up carbon monoxide at the lowest rates and tropical soils at the highest rates. Agricultural soils had a lower carbon monoxide uptake than uncultivated soils, presumably because they had less organic matter in the surface layer. There was also some evidence that soils exposed to higher carbon monoxide concentrations (such as near a free-

*The mixing ratio is defined as the fractional volume (mass) of gas in unit volume (mass) of air under the same conditions of temperature and pressure.

way interchange) had a greater carbon monoxide uptake rate. This was attributed to the development of a larger or more effective population of carbon-monoxide-converting microorganisms.

Data concerning the effect of temperature on the sink capacity of soils are limited and contradictory. Ingersoll *et al.*[206] and Seiler[397] agree that the net uptake rate is drastically reduced as the temperature is increased from 30 C to 50 C. At temperatures in the range of 10 C to 30 C, Seiler[397] measured high uptake rates, whereas Ingersoll *et al.*[206] found the maximum uptake rate to be at 30 C with greatly decreased rates at both 10 C and 20 C.

Seiler (1974)[397] calculated the average soil carbon monoxide uptake rate to be 1.5×10^{-11} g/cm^2/s at 0.2 ppm carbon monoxide. Since the carbon monoxide concentration used in this study is typical of the concentrations of the gas found in the Northern Hemisphere, this value appears to be one of the most reliable estimates of the carbon monoxide uptake rate by soil. The measurement made by Inman *et al.*[208] of 23.4×10^{-11} g/cm^2/s at a carbon monoxide concentration of 100 ppm cannot be safely extrapolated to typical ambient concentrations because the relationship between the carbon monoxide uptake rate and concentration is not known over this range. In highly contaminated areas, where ambient concentrations are particularly high, however, soil carbon monoxide uptake rates could approach this level. The relationship between uptake rate and concentration needs to be determined before uptake in the contaminated areas can be evaluated.

Seiler[397] and Smith *et al.*[412] have shown that production of carbon monoxide also occurs in soils. It is not clear whether this results from nonbiological processes or from a combination of biological and nonbiological processes. Apparently the net rate of carbon monoxide removal by soil is determined by the effects of uptake and production combined. Additional studies, perhaps using carbon-14 (^{14}C), are needed to clarify this point.

To discover which types of soil microorganisms were most active in carbon monoxide uptake, Inman and Ingersoll[207] isolated over 200 different species of fungi, yeast, and bacteria from three soil samples. They found 14 species of fungi capable of removing carbon monoxide from an artificial atmosphere. These included four strains each of *Penicillium digitatum* and *Penicillium restrictum*; four species each of *Aspergillus* and *Mucor hiemalis*; and two strains each of *Haplosporangium parvum* and *Mortierella vesiculata*.

Bacteria also have been isolated from soils that utilize carbon monoxide in metabolism or fermentation. Kluyver and Schnellen[235] reported that under anaerobic conditions the species *Methanosarcina barkerii* produces methane, utilizing carbon monoxide as the only source of carbon. They presented evidence that this fermentation proceeds in the following two steps:

$$4CO + 4H_2O \rightarrow 4CO_2 + rH_2 \qquad (61)$$

$$CO_2 + 4H_2 \rightarrow CH_4 + 2H_2O \qquad (62)$$

Combining these two equations gives the following net reaction:

$$4CO + 2H_2O \rightarrow 3CO_2 + CH_4 \qquad (63)$$

Evidence for the existence of an autotrophic aerobic carbon-monoxide-utilizing bacteria is not conclusive,[360] although several isolated species have been reported to have the capacity to oxidize carbon monoxide to carbon dioxide. Since many of these organisms also could oxidize hydrogen or assimilate simple organic compounds, their true aerobic and autotrophic nature has been questioned.[100]

Vegetation

Some plant species apparently have the ability to remove carbon monoxide from the atmosphere, but the available data are conflicting. In 1957, Krall and Tolbert[239] exposed excised barley leaves to an artificial atmosphere containing 60% carbon monoxide-^{14}C (^{14}CO) and determined the rate of conversion to serine and other compounds. The conversion at 28 C was 0.038 μmol/g/hr. In 1972, Bidwell and Fraser[49] investigated the incorporation of carbon-14 (^{14}C) from an artificial carbon monoxide atmosphere into plant carbon compounds (mainly sucrose and serine). The uptake of carbon monoxide by excised leaves was found in seven out of the nine species studied. Six species showed a measurable uptake rate at concentrations of 2 ppm or lower. If we assume that the uptake rate varies linearly with the concentration in the range 0.2 to 2 ppm, the data for these six species indicate an average uptake rate of 0.003 $\mu mol/dm^2/hr$ at 0.2 ppm. This would be equivalent to $0.23 \times 10^{-11}\ g/cm^2/s$ if we assume a leaf area that is 10 times the ground surface area. This is 15% of the average value reported by Seiler[397] for soils at this concentration, indicating that soils are probably more important than plants as a sink for carbon monoxide. In 1973, Kortschak and Nickell,[237] using methods similar to those of Bidwell, found the uptake of carbon monoxide by the leaves of sugarcane to be $10^{-4}\ mg/cm^2/hr$ at 2 ppm carbon monoxide. In contrast, Inman and Ingersoll[207] were unable to measure any carbon monoxide removal from an artificial atmosphere of 100 ppm by any of the 15 species of higher plants they tested, although their methods were more than adequate to establish carbon monoxide uptake rates by different soils. The assessment of the vegetation as a sink for carbon monoxide will have to wait until this discrepancy is resolved.

Several workers such as Delwiche[120] demonstrated the production of carbon monoxide by higher plants. The relative importance of plants as a source and as a sink for carbon monoxide cannot be determined from the available data.

The uptake rate of carbon monoxide needs to be determined for a number of plant species over the range of atmospheric carbon monoxide concentrations found in ambient and polluted atmospheres. Also, the rates of evolution of carbon monoxide by plants, if any, need to be known before the overall sink properties can be ascertained. In addition, the relationship between soil carbon monoxide uptake and carbon monoxide concentration should be determined, especially in the range between 0.2 ppm and 50 ppm. The effects of temperature on the evolution of carbon monoxide by the soil also still remain to be learned. Knowledge of these parameters will improve the determination of the carbon monoxide sink properties of soils and vegetation.

Biochemical Removal

The binding of carbon monoxide by porphyrin-type compounds, found in plants and animals, is analogous to carbon monoxide uptake by hemoglobin in blood and is a potential sink for carbon monoxide. However, the carbon monoxide is reversibly bound by the heme compounds found in man and animals and thus is eventually discharged from the blood, and only a small fraction of carbon monoxide is retained.[282]

Absorption in the Oceans

Unlike carbon dioxide, the oceans can no longer be considered a sink for atmospheric carbon monoxide.[219] In fact, in the geographical areas studied, the carbon monoxide supersaturation and the high diurnal variations of carbon monoxide observed in the upper layer of the ocean provide evidence that the oceans are a significant source of carbon monoxide that comes from the photobiologic processes of marine algae. Inland expanses of fresh water have not been adequately assessed at present, but it is doubtful that they can serve as a sink for the large amount of carbon monoxide injected into the atmosphere.

Removal at Surfaces

Kummler *et al.*[246] have extrapolated the reported heterogeneous reaction rates of nitrous oxide with carbon monoxide to ambient temperatures (300 K) and suggest that atmospheric reaction between these gases in the presence of such common materials as charcoal, carbon black, or glass is a feasible scavenging process for atmospheric carbon monoxide. This mechanism is derived from observations made by Gardner and Petrucci[149] of the chemisorption of carbon monoxide at room temperature on the oxide films of copper, cobalt, and nickel using infrared spectroscopy. Also, Liberti[260] has observed from analyses of collected urban particulate material that quantities of carbon monoxide ranging from 10 to 30 $\mu g/g$ were associated with the aerosol. He suggested that absorption by dust may be an important removal mechanism for ambient carbon monoxide and, through co-deposition, it is made available to soil microorganisms for oxidation. These studies are inconclusive, and additional investigations are necessary. Quantitative evaluation of the catalytic efficiency and absorbing capacity of common surfaces and atmospheric particulate materials under ambient conditions is required to determine whether or not they provide adequate sinks for atmospheric carbon monoxide.

4

Environmental Analysis and Monitoring

There are major problems in correlating atmospheric data with health effects data for carbon monoxide pollution despite both the financial investments in monitoring and legal reliance placed on these data. Some of the problems related to both the nature of the pollutant and the monitoring devices used will be considered in this chapter.

There are varying amounts of a large number of trace substances in the atmosphere. When the concentration of certain of these increases above a threshold amount, there is a harmful effect on human health or welfare. When this happens, the atmosphere is considered polluted, and the causal agents are called air pollutants. Most substances considered to be air pollutants originate from both natural and man-made sources; therefore, origin cannot be used as a criterion for distinguishing between pollutants and nonpollutants.

The determination of the threshold concentration for each pollutant is difficult and very often controversial. Therefore, specification of a polluted atmosphere is based upon concentration standards established by a consensus among panels of experts designated by air pollution control officials. A distinction is made between primary standards, which are based on health effects, and secondary standards, which relate to welfare.

The amount of carbon monoxide emitted into the atmosphere is about

10 times greater from natural than from anthropogenic sources. However, the natural sources are so widely dispersed that their contribution to atmospheric pollution can be neglected. Analyzing pollutant exposure is complicated both because peoples' activity patterns are unpredictable and because there is a wide concentration variation with respect to location and time for carbon monoxide from anthropogenic sources.

Carbon monoxide, when taken into the body, is converted into carboxyhemoglobin and then quantitatively eliminated. In a uniform environment, its concentration in the body reaches a steady state and then remains constant. Exposure to an environment with a higher carbon monoxide concentration increases the body burden, while exposure to an environment with a lower concentration causes the elimination of carbon monoxide, thus decreasing the total body burden. Such changes characteristically take place over several hours.

Another consideration is the choice of the best way to express the large number of measurements. Conventionally, the "hourly mean" values and "8-hr means" have been expressed as arithmetic means. The "hourly mean" and the "8-hr mean" were calculated on the assumption that a significant number of people will be exposed either for 1 or 8 hr. In principle, the measure of central tendency used should be such that those periods having the same mean value should also have the same physiologic manifestations. For any averaging process, however, there are limiting cases in which this condition cannot be met; for example, exposure to a low concentration for 8 hr is not equivalent to a 5-min exposure to a lethal concentration followed by 7 hr and 55 min at zero concentration. But even in less extreme cases, exposure periods with the same arithmetic average concentration do not all stress the receptor equally. Additional study is needed to arrive at the optimal expression of central tendency. When a large number of hourly arithmetic means have been collected, it is frequently found that they have a lognormal distribution, which permits the expression of the entire data distribution as a geometric mean and a geometric standard deviation, i.e., the entire ensemble of data can be represented by two numbers. This compactness of expression is very convenient, and it is not intrinsically incorrect. Nevertheless, the limitations of the data must be kept in mind so that the researcher does not do further statistical analyses that ignore the character of the data. For example, when using these statistical data to compute the optimum number of sampling stations within a monitoring network and the optimum sampling frequency, the individual sample data taken at sequential times and adjacent locations should be checked for correlation. It is possible to reduce the number of sampling stations if good correlation can be shown. Furthermore, the data on each pollutant must be evaluated separately owing to differences in source–receptor relationships that affect the correlations for pollutants other than carbon monoxide.

To summarize, it is convenient to use statistical measures of central tendency and dispersion to represent aerometric data; these measurements are so closely correlated in space and time that it is disadvantageous to use additional statistical manipulations that depend on the randomness or independence of successive measurements. As yet, the ideal spacing of monitors has not been achieved.

EMISSION OF CARBON MONOXIDE

Motor vehicles are the largest anthropogenic source of atmospheric carbon monoxide. Diesel-powered vehicles emit a much lower amount than those using gasoline as a fuel. A 1972 gasoline-powered light-duty vehicle emits 59.0 g/km (36.9 g/m) as compared to a pre-1973 diesel-powered light-duty vehicle that emits 2.7 g/km (1.7 g/m).[452] A compilation of the estimated man-made emissions of the major pollutants for the United States and for three major cities is given in Table 4-1. The breakdown of carbon monoxide emissions by source, both nationally and for New York City, given in Table 4-2, shows the preponderance in the mobile sources.

In individual locations, however, stationary sources can be equally important. For example, proximity to a poorly controlled petroleum refinery could, under certain circumstances, result in exposures to significant concentrations of carbon monoxide.

Emission factors for a number of sources are listed in Table 4-3. The industrial factors are multiplied by appropriate measures of industry

TABLE 4-1 Estimated Man-Made Emissions for 1973

	Emissions, 10^3 tons/yr				
City	Sulfur Dioxide	Particulates	Nitrogen Oxides	Hydro-carbons	Carbon Monoxide
U.S. total (1972)[a]	33,210	19,800	24,640	27,820	107,301
New York City[b]	131	47	317	197	495
Los Angeles[c]	133	47	407	281	2,664
Chicago[d]	99	74	112	93	364

[a]U.S. Environmental Protection Agency.[456]
[b]New York City Department of Air Resources.[82]
[c]Los Angeles County Air Pollution Control District.[280]
[d]City of Chicago Department of Environmental Control.[81]

TABLE 4-2 Carbon Monoxide Emissions

	Emissions, 10^3 tons/yr			
Source	U.S., 1972[a]	Percent of Total	New York City, 1973[b]	Percent of Total
Stationary	24,276	23.9	28	5.6
Fuel combustion	1,180	1.2	24	4.8
Coal	772	0.8	18	3.6
Fuel oil	96	0.1	4	0.8
Natural gas	171	0.2	2	0.4
Wood	49	—	—	—
Other	92	0.1	—	—
Industrial processes	17,469	17.2	trace	0.1
Solid-waste disposal	4,982	4.9	4	0.7
Miscellaneous	645	0.6	—	—
Mobile sources	77,288	76.1	469	94.4
Motor vehicles	69,560	68.4	455	91.5
Gasoline	68,850	67.7	440	88.5
Diesel	710	0.7	15	3.0
Aircraft	976	1.0	14	2.8
Vessels	437	0.4	—	—
Railroads	149	0.1	—	—
Other (off-highway)	6,296	6.2	—	—
Total	101,564	100.0	497	100.0

[a] U.S. Environmental Protection Agency.[456]
[b] New York City Department of Air Resources.[83]

size to obtain the total emissions.[452] To obtain the total emissions for motor vehicles, the emission factor adjusted by the appropriate speed correction factor is multiplied by the total vehicle miles traveled. Figure 4-1 has been calculated from data in the reference and gives the speed correction factor for 1975 with the additional assumption that 88% were autos and 12% were light-duty trucks.[453]

Vehicle speed has long been recognized as a critical variable in predicting motor vehicle emissions. It is only recently with the introduction of sophisticated emission control systems such as catalysts and stratified charge systems that the importance of ambient temperature and hot/cold weighting (a measure of the relative mileage contribution of warmed-up vehicles) has been recognized. This has led to the quantification of the corresponding emission adjustment factors. Calculations based upon USEPA's emission factor report indicate that carbon monoxide emissions

TABLE 4-3 Selected Carbon Monoxide Emission Factors[452]

Activity	Conditions	Factor
Bituminous coal combustion	Utility and large industrial boilers	1 lb/ton coal (0.5 kg/ton)
Combustion	Large commercial and general industrial boilers	2 lb/ton coal (1.0 kg/ton)
Combustion	Commercial and domestic furnaces	10 lb/ton coal (5.0 kg/ton)
Combustion	Hand-fired unit	90 lb/ton coal (45 kg/ton)
Fuel oil combustion	Power plants	3 lb/10^3 gal (0.36 kg/10^3 l)
	Industrial, commercial, and domestic	4–5 lb/10^3 gal (0.47–0.6 kg/10^3 l)
Carbon black manufacturing	Channel process	33,500 lb/ton (16,750 kg/ton)
	Thermal process	Negligible
	Furnace process	5,000 lb/ton (2,500 kg/ton)
Charcoal manufacture	Pyrolysis of wood	320 lb/ton (160 kg/ton)
Meat smokehouses	—	0.6 lb/ton meat (0.3 kg/ton)
Sugarcane processing	Field burning	225 lb/acre burned (253 kg/ha)
Coke manufacture	Without controls	0.6 lb/ton (0.3 kg/ton)
Steel mills	Uncontrolled	1,750 lb/ton (875 kg/ton)
Foundries	Gray iron cupola	145 lb/ton (72.5 kg/ton)
Petroleum refineries	Uncontrolled fluid catalytic cracking	13,700 lb/10^3 bbl (38.63 kg/10^3 l)
	Uncontrolled moving bed cracking	3,800 lb/10^3 bbl (10.7 kg/10^3 l)
Highways[a]		
1965	Approximately 20 mph (32 km/hr) for each of the years	89 g/mi (55.3 g/km)
1970		78 g/mi (48.5 g/km)
1972		76.5 g/mi (47.6 g/km)
1973		71.5 g/mi (44.4 g/km)
1974		67.5 g/mi (40.6 g/km)
1975		61.1 g/mi (38.0 g/km)
1980		31.0 g/mi (19.3 g/km)
1990		11.3 g/mi (7.02 g/km)

[a]Average emission factors for all vehicles on the road in given year.

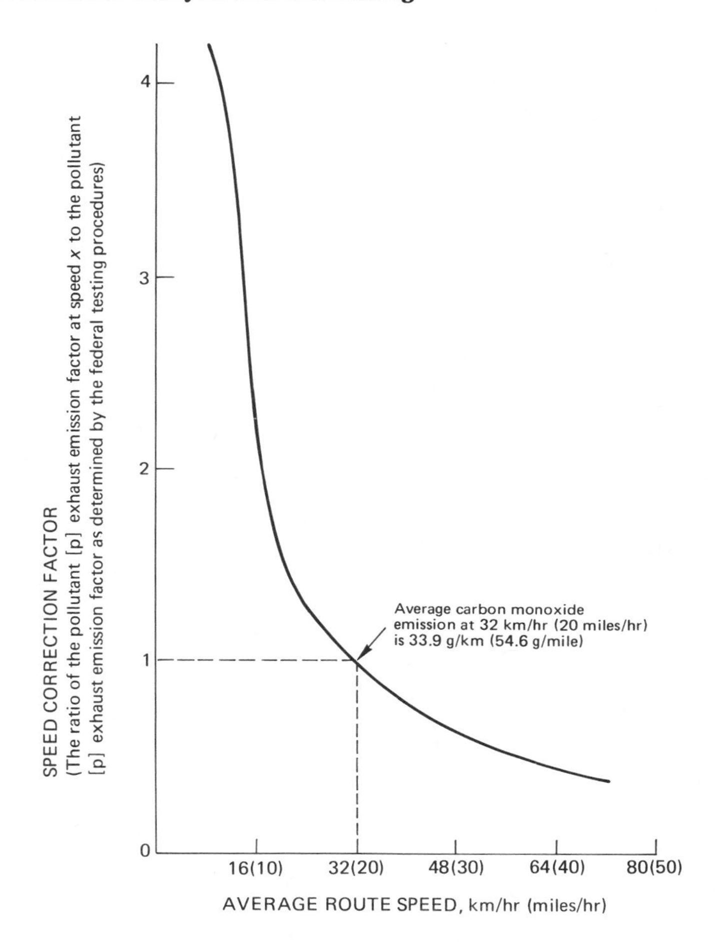

FIGURE 4-1 The relationship between emissions at a given mean traffic speed and emissions at 32 km/hr (20 mi/hr). Based upon the model year mix existing in 1975, an automobile to light-duty truck ratio of 8.8 to 12 is assumed.

from conventional vehicles not equipped with catalytic exhaust control devices are three times greater during cold starts at −7 C (20 F) than at 27 C (80 F).[452] Over this same ambient temperature range the carbon monoxide emissions of catalyst-equipped vehicles during cold starts differ 10-fold. As these improved pollution-control systems become more prevalent, it will be particularly necessary to take adjustment factors into consideration.

It is often difficult to obtain a sufficient number of measurements on which to base the calculation of emission factors. This is especially true for sources with highly variable emissions since experience has shown that sizable errors can be introduced by assuming that the measurements from such sources are an accurate representation.

The emission factors for motor vehicles are much better documented because vehicle emissions have been extensively and carefully measured.

The amount of carbon monoxide from natural sources is about 10 times the amount from anthropogenic sources. The largest source is the oxidation of atmospheric methane. There is also a marine biologic source and lesser contributions from the decomposition of vegetation and from forest fires.[472] With the exception of natural combustion sources, all of the other sources are widely dispersed and result in rather uniformly low concentrations worldwide.

AMBIENT CONCENTRATIONS

In contrast to natural emissions, anthropogenic emissions are invariably concentrated. While individual plants may discharge large amounts of carbon monoxide, these are usually located in fairly open terrain. The highest concentrations experienced by a significant portion of the population tend to be caused by vehicular traffic in the narrow and poorly ventilated street canyons of major cities, since ambient concentrations are the result of both mass discharge and degree of dilution.

OBSERVED CONCENTRATIONS

Range of Measurements

An idea of the range of carbon monoxide concentrations experienced in urban centers is given in Figure 4-2, which shows the distribution of hourly average concentrations for a number of cities on a logarithmic probability plot.[456] The approximate similarity in the slopes indicates a narrow range (in the vicinity of 2.0) of geometric standard deviations. There is, however, nearly a 10-fold range in the median values, which would be even greater if smaller cities were included. These measurements should not be construed as representative of the entire city in question, since they were taken at a single station that may not have been ideally located.

When samples of carbon monoxide are taken within or above urban canyonlike streets, marked differences in the range of measured concentrations are observed. Tables 4-4 and 4-5 summarize measurements made at the pedestrian level and at the rooftop level both in densely

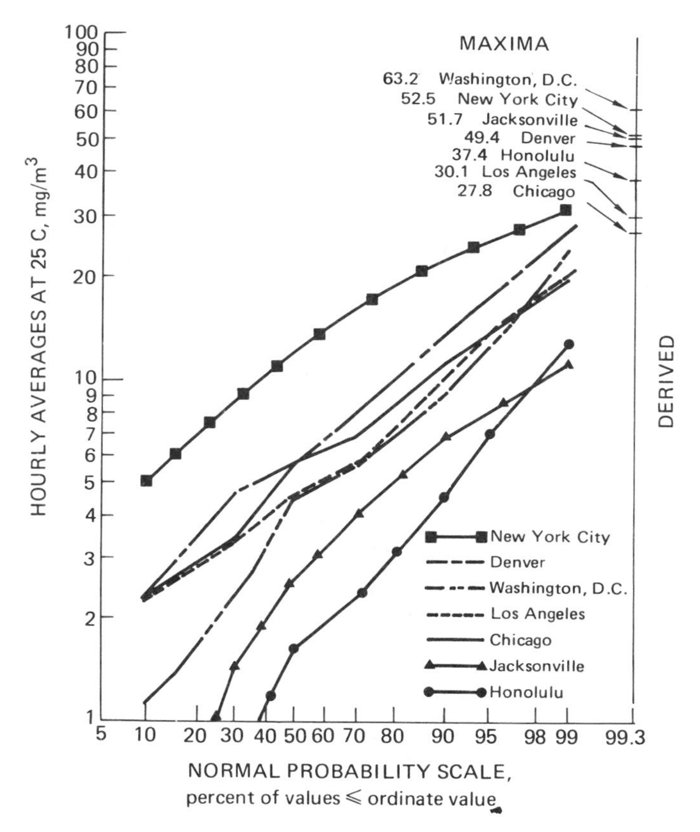

FIGURE 4-2 Carbon monoxide distribution of 1-hr averages, 1972. (From U.S. Environmental Protection Agency.[456])

populated and in heavily trafficked areas of New York City. The heights of the rooftop stations range from 7.5–29.9 m (25–100 ft) above street level, and the concentrations at these heights are indicative of those at rooftops and near the street level at locations slightly removed from heavy traffic. In following their daily life patterns, a large number of people are exposed to the time-averaged concentrations shown in Table 4-4 (at street level) or to those shown in Table 4-5 (at rooftop level). Most people, however, are exposed to some combination of these concentrations. It appears that these two sets of data would lead to completely different evaluations of the carbon monoxide problem.

TABLE 4-4 Carbon Monoxide (ppm) Hourly Readings, Street Level Stations, 1974 Calendar Year[a]

									8 hr (7 a.m.-3 p.m. EST)							
	No. of Readings		Arithmetic Average		Peak		No. of Hours > 35		No. of 8-hr Averages		Average		Peak		No. > 9	
Station	1973	1974	1973	1974	1973	1974	1973	1974	1973	1974	1973	1974	1973	1974	1973	1974
00—121 Street Laboratory	6,018	5,473	4.5	4.1	35	26	0	0	213	192	4.5	4.5	13.0	15.6	8	5
94—59th Street Bridge	7,140	8,463	18.5	16.2	70	57	384	89	277	335	21.7	18.5	49.6	37.4	276	334
96—Canal Street	7,635	7,698	8.8	9.3	36	45	1	1	302	309	10.4	11.0	21.5	20.5	184	217
98—45th Street and Lexington Avenue	8,503	8,698	11.3	12.6	61	43	29	5	349	355	13.6	14.6	25.3	25.8	264	307

[a] City of New York, Bureau of Technical Services, Department of Air Resources.[82]

TABLE 4-5 Carbon Monoxide (ppm) Hourly Readings, Rooftop Stations, 1974 Calendar Year[a]

							8 hr (7 a.m.-3 p.m. EST)							
	No. of Readings		Arithmetic Average		Peak		No. of 8-hr Averages		Average		Peak		No. > 9	
Station	1973	1974	1973	1974	1973	1974	1973	1974	1973	1974	1973	1974	1973	1974
Citywide	68,280	59,443	3.7	3.3	34	23	2,782	2,406	3.8	3.4	15	12	24	6
1—Bronx High School of Science	7,079	7,069	3.6	4.2	21	19	342	314	3.6	4.2	9	8	2	0
3—Morrisania	6,381	7,081	3.4	2.6	20	13	292	288	3.6	2.6	12	10	3	1
14—Queens College	7,657	7,755	3.8	3.5	20	19	325	311	4.0	3.5	11	8	2	0
30—Springfield Gardens	5,759	1,889	5.2	4.3	28	23	184	77	5.3	4.6	11	12	6	2
5—Central Park Arsenal	6,866	5,673	4.1	2.9	34	15	289	224	3.8	3.0	14	10	4	1
10—Mabel Dean Bacon	6,708	5,543	3.6	3.1	16	13	232	214	3.9	3.3	10	10	2	1
11—Greenpoint	8,210	7,329	3.7	3.6	22	18	266	293	3.8	3.5	15	10	3	1
18—Brooklyn Public Library	6,720	2,843	3.4	3.7	20	18	241	117	3.8	3.9	14	8	1	0
26—Sheepshead Bay High School	5,521	7,408	3.3	2.5	20	20	283	292	3.5	2.6	14	8	1	0
34—Seaview Hospital	7,379	6,853	2.9	2.5	11	8	328	276	3.2	2.6	9	6	0	0

[a]City of New York, Bureau of Technical Services, Department of Air Resources.[82]

Temporal Variations

A lognormal plot (normalized logarithmic probability plot) of data separated both by season and by day of the week[215] for a station in New York City is shown in Figure 4-3. Seasonal differences are small compared with the difference between Sundays and weekdays. This is probably not surprising, since there is little seasonal difference in New York City's traffic density. Sunday traffic, however, is significantly lighter than weekday traffic, and this correlation shows up clearly.[215,216]

The correlation between carbon monoxide concentration and traffic is still more apparent in Figure 4-4.[103] These plots of the diurnal course of both traffic and carbon monoxide reveal their typical patterns and show their similarity. Cities with a marked rush-hour traffic peak in the morning and afternoon tend to show similar carbon monoxide patterns. Cities such as New York that experience saturation traffic throughout business hours tend to show a plateau during the day rather than a midday minimum.

Spatial Variations

In narrow canyonlike streets, most of the carbon monoxide is emitted at ground level. Dilution is largely through mechanical turbulence in-

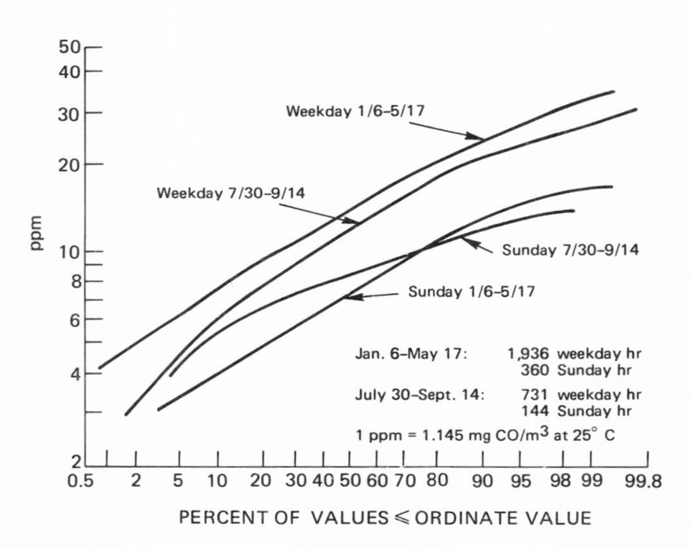

FIGURE 4-3 Cumulative distributions (1967) of hourly average concentrations of carbon monoxide (110 East 45th Street, Manhattan). (From Johnson *et al.*[215])

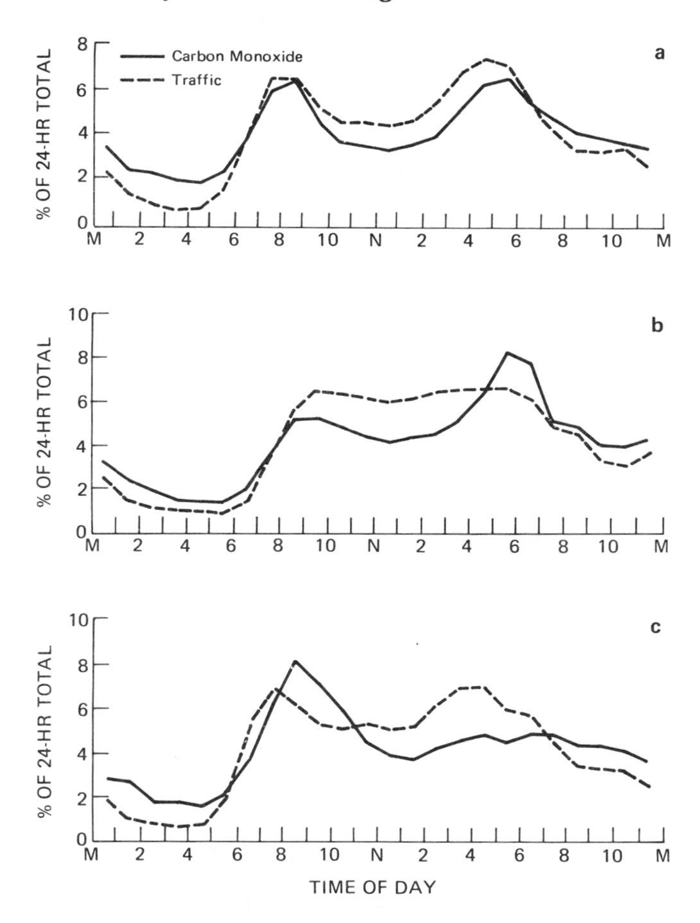

FIGURE 4-4 Diurnal variation of carbon monoxide and traffic. (a) Lodge-Ford Freeway interchange (correlation coefficient, 0.92); (b) Columbus Circle (correlation coefficient, 0.86); (c) Harbor-Santa Monica Freeway interchange (correlation coefficient, 0.75). (Reprinted with permission from Colucci and Begeman [103] [J. M. Colucci and C. R. Begeman, Carbon monoxide in Detroit, New York, and Los Angeles air, Environ. Sci. Technol. 3:41-47, 1969. Copyright by the American Chemical Society].)

duced by the movement of the traffic and possibly by convection from the excess heat produced by the vehicles. The resulting vertical distribution (Figure 4-5) shows smoothed vertical profiles up the sides of two high-rise buildings in New York City.[151] The data are given by season to distinguish between the heating and nonheating seasons. The differences shown are probably not significant. One conclusion that can be drawn is that the concentrations measured at a given station are very sensitive to the height of the intake tube. This point is discussed further below.

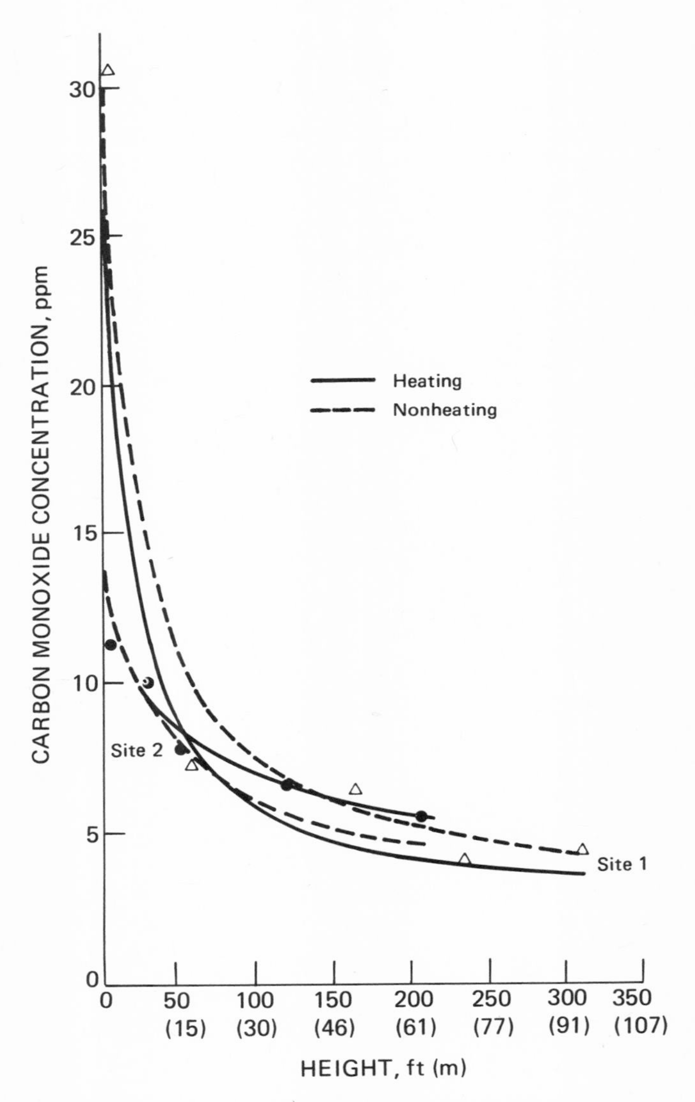

FIGURE 4-5 Vertical outdoor carbon monoxide profile—both sites. (From General Electric Company.[151])

Meteorological Effects

An additional complication occurs when a prevailing wind blows across a deep street canyon. Studies have shown that this induces a downward air motion on the side of the street facing the wind, a flow across the street in the opposite direction to the prevailing wind, and an upflow on the upwind side of the street. This reverse eddy causes low carbon monoxide concentrations on the downwind side of the street and high concentrations on the upwind side.[328]

Local wind behavior combined with the configurations of nearby structures can be responsible for significant differences in the measurements reported by closely adjacent monitors. For example, differences as great as twofold were found at different corners of the same intersection in Tokyo.[445]

Indoor-Outdoor Relationships

Because carbon monoxide is relatively inert, there is little adsorption on surfaces. Therefore, with the possible exception of a brief time lag, a close relationship would be expected between concentrations in the open air and those inside buildings. Figure 4-6, which gives indoor and outdoor concentrations and adjacent traffic count for a third-floor apartment in a high-rise building in New York City, shows that this is true in many cases.[151] The daily course of carbon monoxide for a detached suburban home (Figure 4-7), where leakage from an attached garage and the effects of gas cooking and smoking largely obscure the outdoor influences, is in marked contrast.[45] The effects of indoor sources, including tobacco smoking, predominate over the influence of the outside atmosphere. Indoor-outdoor combined effects will be increased if windows and doors are open and decreased if the house is tightly closed, as is the case during the heating season.

Summary

Urban carbon monoxide concentrations are at least 10 times higher than background concentrations; seasonal differences are small; concentrations on Sundays are almost invariably less than those on weekdays; diurnal concentration patterns follow diurnal traffic patterns, with a tendency to peak during morning and evening rush hours (the intervening period may or may not show a decrease, depending on the nature of nearby traffic); concentrations decrease steeply with increasing height and are also affected, even when averaged over the long term, by persis-

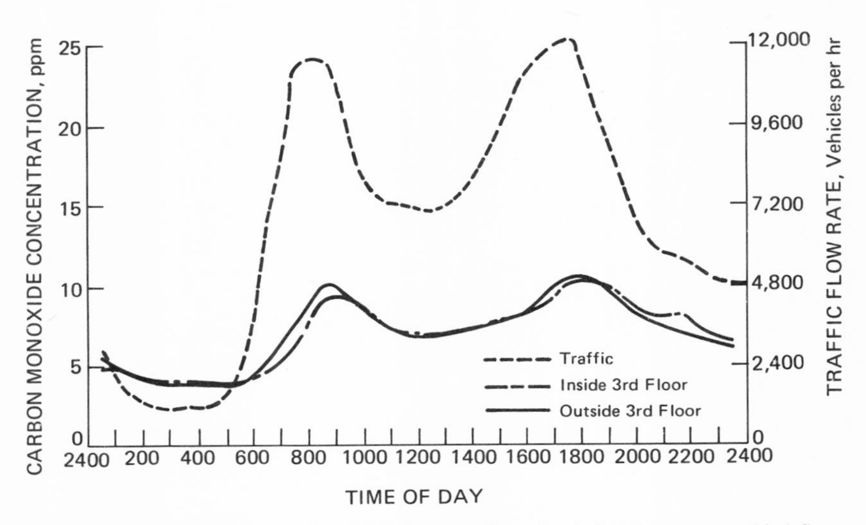

FIGURE 4-6 Diurnal carbon monoxide and traffic—site 1, heating season, third floor, weekdays. (From General Electric Company.[151])

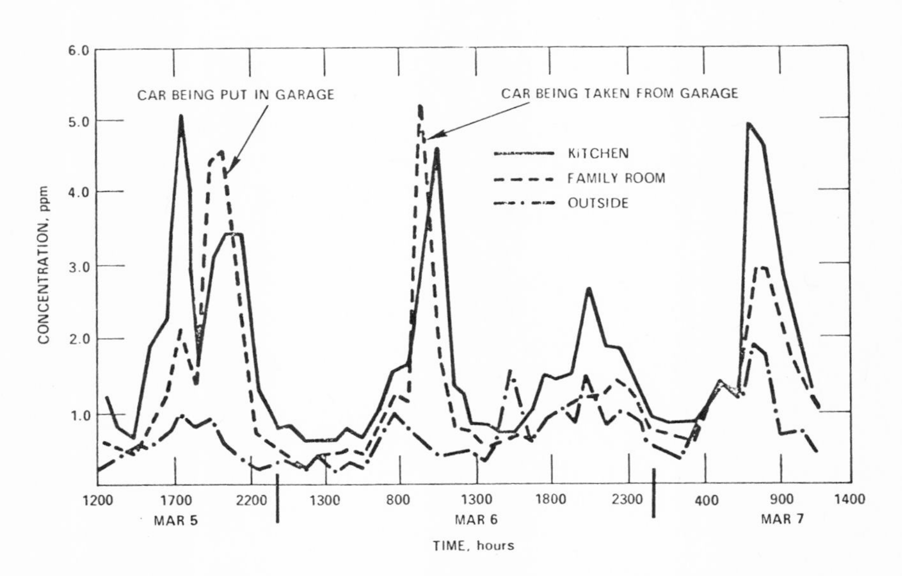

FIGURE 4-7 Indoor-outdoor diurnal variation in carbon monoxide. (From Benson *et al.*[45])

tent air circulation patterns. Because carbon monoxide emission decreases and mechanical turbulence increases with increasing vehicle speed, high-speed roads tend to yield lower ambient carbon monoxide concentrations in their vicinity, even though the traffic density is increased. In addition, because high-speed roads tend to be built in open spaces and on elevated roadbeds, they are less subject to canyon effects.

MEASUREMENT OF CARBON MONOXIDE

Several of the techniques used for monitoring carbon monoxide are discussed in detail in Appendix A. The previous discussion has pointed out a number of measurement problems common to all monitoring techniques. There are, however, additional problems specific to the measurement of carbon monoxide. The Environmental Protection Agency has set ambient air quality standards for carbon monoxide based on the resulting concentrations of carboxyhemoglobin in human blood. These were supposed to remain below 2% in nonsmokers, which corresponds to carbon monoxide concentrations of 40 mg/m^3 (35 ppm) for 1 hr or 10 mg/m^3 (9 ppm) for 8 hr.[454] Since there is a finite probability of exceeding any given concentration, the statistical status of these standards was further defined by stating that they could be exceeded by experiencing one higher concentration per year without constituting a violation. That is to say, these standards refer to the second highest concentration measured in any year. It has recently been argued that using three fixed periods per day (0000–0800, 0800–1600, and 1600–2400) for the 8-hr averages leads to underestimating concentrations. Therefore, 8-hr moving averages of 1-hr average concentrations should be used instead.[304] The number of violations that result if running averages are used should be limited by counting only nonoverlapping 8-hr averages as violations.

These standards have had a profound effect on both the measurement and reporting of carbon monoxide concentrations. Whether or not the selection of concentrations has been completely accurate, it can be seen from the foregoing discussion that a typical mobile individual may experience a history of carbon monoxide exposure very different from that recorded by either one or several stationary monitoring devices. A situation can be pictured in which a monitor at the second-floor level would measure one concentration, while a person on the curb directly beneath would experience 10 times that concentration, and another person on an upper floor would be exposed to significantly less than the measured concentration. A sampling probe can always be located so that much higher concentrations are measured than a person in the vicinity could reasonably experience for any extended period.

The statistical problem of the optimal theoretical design for sampling systems has not yet been solved. Consequently, to obtain reliable data, first the number and locations of the sampling stations must be decided and then many trials must be made to minimize any distortion of the data because of one or two anomalous locations.

Assuming optimal locations have been selected and tested, the quality control of the individual monitoring instruments remains a problem. All carbon-monoxide-monitoring instruments have features in common: they contain complex electronic circuits, they are operated near the sensitivity limits set by inherent noise, and, most significant of all, their calibration is arbitrary. None of the available instruments can be calibrated solely from consideration of its operating principle. For this reason the instruments must be calibrated using standard gas mixtures. Their accuracy is thus limited by the accuracy of available standards. Additional potential sources of error are the electronic circuits that add the possibility of malfunction or calibration drift not necessarily obvious from inspection of the output data and unless a program of frequent calibration, preferably daily, is instituted. Much of the carbon monoxide monitoring in the past has been performed by nondispersive infrared instruments, which are very susceptible to water vapor and which require highly trained technicians. Thus, some of the data generated have been unreliable.

SIMULATION MODELING

The preceding discussion has pointed out the inherent difficulties in obtaining carbon monoxide concentration data in order to calculate accurately human exposures. Ideally, the individuals being tested should carry a personnel monitor that would continuously record the concentrations to which they are exposed. Alternatively, a detailed record of their movements combined with monitoring data representative of each location could be used to calculate exposures and rate of carbon monoxide uptake. For technical and economic reasons, neither of these alternatives is feasible at present.

A major step towards the ability to characterize the concentration in a complex area has been the development of simulation models. Once a successful model has been achieved, estimates of concentrations at a particular location can be made, given the intensity of emissions and their spatial and temporal patterns, the micrometeorology, and the structural configurations bounding the area. In combination with a relatively small number of strategically located monitoring stations, such models can be used to depict the concentrations at many locations in a large area. Unavailability of such sophistication in the past necessitated using oversimplified models. The pressures created by the cost of pollution control require investing in the development of more effective models. Modeling has also become important in the preparation of environmental impact statements. Therefore, there is an increased effort to develop successful air-quality models. The models for pollutants such as sulfur dioxide are usually long-term. Because of the difficulty in depicting meteorological conditions over extended periods, these can be in error by as much as 100% or more.

Short-term models for urban carbon monoxide conditions are currently under development in order to meet state implementation plan requirements and to prepare environmental impact statements. These short-term models are capable of calculating concentrations within 25% of measured values.[328] Because of the short-term relation between carbon monoxide exposure and health effects, concentration models have primarily emphasized hourly averages.

MONITORING CARBON MONOXIDE IN THE ENVIRONMENT BY BIOLOGIC TECHNIQUES

Carbon monoxide is not significantly altered metabolically when taken into the body. It binds reversibly with heme pigments, principally hemoglobin, the red pigment of the blood. A direct indication of the amount of human exposure can therefore be obtained by hemoglobin measurements. From the standpoint of health, these are more significant than ambient air measurements, since the concentration of carboxyhemoglobin (HbCO) in the blood is related to the physiologic effects of carbon monoxide on people.

There are practical limits to the value of biologic sampling for carbon monoxide uptake. Carbon monoxide is taken up relatively slowly by the body from ambient sources. Since the rate of uptake is dependent on several physiologic factors, the interpretation of measurements in terms of ambient sources may not be simple. Because cigarette smoking is a major source of carbon monoxide exposure, widely prevalent in the urban population, ambient air sources can be evaluated only in nonsmokers. Occupational sources of carbon monoxide exposure, for example in garages, may also contribute to its presence in urban dwellers.[80,119,364]

Routine sampling of blood has two additional practical limitations; the first is that the discomfort of taking a blood sample by any method would make random sampling in the general population unacceptable, and the second is that analytical techniques for determining low concentrations of carboxyhemoglobin are as yet not sufficiently reliable. At the time of the previous National Academy of Sciences report (1969), rapid spectrophotometric methods were not considered accurate for measuring very low carboxyhemoglobin concentrations.[326] Accuracy is crucial if the blood carboxyhemoglobin is used to indicate exposure to the low concentrations in urban air. The methods currently used to measure carboxyhemoglobin are described in Appendix A.

Carboxyhemoglobin can also be estimated indirectly by measuring alveolar gas, which is the gas present in the deep regions of the lungs. Using subjects with normal lungs and properly applying this technique, good results are achieved with a minimum of discomfort. The subject's cooperation and well-trained personnel are required. Therefore, the wide applicability of this technique for routine sampling of carbon monoxide exposure is unlikely. This technique is described in Appendix B. Because a small amount of carbon monoxide is generated within the body, there is about 0.4% carboxyhemoglobin present even when no carbon monoxide is in the inspired air, which adds a further complication to biologic sampling for carbon monoxide exposure. The production rate of carboxyhemoglobin can be increased by certain diseases and physiologic conditions, such as hemolytic anemias.[95] These can increase the carboxyhemoglobin 2 to 3% above normal endogenous production. It is also increased by ingesting drugs that induce the hepatic-oxidizing enzyme cytochrome P-450.[88] For example, barbiturates slightly increase carbon monoxide production.[88]

We can add the effects of carbon monoxide produced in the body to that of the carbon monoxide from the ambient air.[91] The expected carboxyhemoglobin concentration at equilibrium, calculated using an endogenous carbon monoxide production rate of 0.4 ml/hr, is shown in Table 4-6. For comparison, there is a calculation based on the empirical equation: % carboxyhemoglobin $= 0.4 + (p/7)$, where p is the inspired carbon monoxide concentration in ppm. Table 4-6 shows that if equilibrium is achieved (for comment see the section on uptake), 2% carboxyhemoglobin should result from inspired carbon monoxide of about 12 ppm, and 3% carboxyhemoglobin should result from about 18 ppm.

The most extensive study of carboxyhemoglobin in the general public has been carried out by Stewart and his associates, who sampled blood drawn from about 31,000 individuals by blood donor mobile units in 17 urban areas and also in some small towns in New Hampshire and Vermont.[420] Their results showed that smokers' blood contained a much higher percent of carboxyhemoglobin than that of nonsmokers. They found, however, that in all regions studied a large percentage of the nonsmokers had over 1.5% carboxyhemoglobin, indicating significant exposure. Frequency distributions for each of these regions indicated that a small percentage (1–2%) of the nonsmokers, despite claiming to

TABLE 4-6 Calculated Carboxyhemoglobin at Equilibrium with Inspired Carbon Monoxide Concentration (Applicable to Nonsmokers Only)

Inspired Carbon Monoxide, ppm	% Carboxyhemoglobin from Coburn *et al.*[a]	% Carboxyhemoglobin from Empirical Equation % Carboxyhemoglobin = 04 + $p_tCO/7$[b]
0	0.36	0.4
5	1.11	1.1
8.7	1.66	1.6
10	1.85	1.8
15	2.57	2.5
20	3.29	3.3
30	4.69	4.7
40	6.05	6.1
50	7.36	7.5

[a] Assumptions used in the equation of Coburn *et al.*:[82] The carbon monoxide production = 0.4 ml/hr STPD; the diffusion capacity D_LCO = 20 ml/min/torr; the barometric pressure = 760 torr; the alveolar ventilation = 3,500 ml/min STPD; the Haldane constant = 220;[380] the mean pulmonary capillary oxygen pressure (p_CO_2) = 100 torr; and the fraction of unbound hemoglobin is constant at 3%.
[b] p_tCO is the inspired carbon monoxide concentration in parts per million. This equation is applicable up to 50 ppm.

be ex-smokers, were actually still smoking. Their inclusion as nonsmokers does not change the results significantly. In addition, some of the high values found in nonsmokers in four of the cities may have been due to special occupations in which unusual exposures to high carbon monoxide concentrations occurred in enclosed areas.[419]

The effect of occupationally related carbon monoxide sources was emphasized by Kahn and associates, who did a similar analytical study of more than 10,000 blood samples from nonsmokers and 6,000 samples from smokers in the St. Louis area.[221] For the nonsmokers classified as "industrial workers" the mean carboxyhemoglobin concentration was 1.4%, and for those classified as "other than industrial" the mean was 0.8%. The overall mean for both employed and unemployed was 0.9%. The conclusions drawn from the results of Kahn *et al.*[221] differ markedly from those of Stewart *et al.*[419,420] Whereas Stewart and his co-workers concluded that 35% of nonsmokers in St. Louis were exposed to ambient carbon monoxide, causing their carboxyhemoglobin to be greater than 1.5%, Kahn *et al.*[221] concluded that only a small percentage of nonsmokers (many of whom were industrial workers) had values greater than 2%. Of the nonindustrial workers, 5.7% had values above 2%, but some of these were unquestionably smokers.

Kahn *et al.*[221] did show that a small urban-rural gradient for carboxyhemoglobin existed in the St. Louis area. Three areas—"highly urban," "urban," and "rural"—were designated according to the 1970 census data. The mean carboxyhemoglobin for nonsmokers who were not industrial workers was 0.8%, 0.7%, and 0.6% for the highly urban, urban, and rural areas, respectively. These very small differences do not support the conclusion that carbon monoxide uptake from ambient sources in the urban air is physiologically significant.

The discrepancy in the conclusions of these two studies is important with respect to current air quality control strategies. Both studies emphasize the significance of smoking and occupational exposures. Stewart's group, however, infers that in cities with such different meteorological conditions and population densities as New York, Los Angeles, Denver, New Orleans, Phoenix, Anchorage, St. Louis, and Washington, one-fourth to three-fourths of the nonsmokers are consistently exposed to carbon monoxide concentrations above the current standard of 8.7 ppm (which would result in carboxyhemoglobin above 1.5%; see Table 4-6). Kahn *et al.*[221] conclude, to the contrary, that the carbon monoxide in the urban environment of St. Louis contributes only negligibly to carbon monoxide uptake by the general population.

Other measurements have been made that apply to the questions raised by these studies. Aronow *et al.*[13,14,16] found carboxyhemoglobin values around 1.0% in 25 nonsmoking subjects in the Los Angeles control area, which rose to 5.1% when these subjects were driven in a car on freeways for 1.5 hr. Ambient carbon monoxide was reported to be 1 to 3 ppm in the laboratory and about 50 ppm in the car. Radford *et al.* (unpublished observations) found values of 0.9% carboxyhemoglobin in the blood of 19 laboratory workers and other personnel in Baltimore who had never smoked, and values of 0.6% carboxyhemoglobin in 32 nonsmoking residents of Hagerstown, Maryland, a small rural city. The carbon monoxide concentrations observed in Baltimore are comparable to those reported in St. Louis, and these results resemble those of the St. Louis study. Horvath (unpublished observations), in Santa Barbara, California, has found 0.6% carboxyhemoglobin in the blood of approximately 150 nonsmokers, indicating low carbon monoxide exposures for this city, even though high ambient air values have been recorded in Santa Barbara. Ayres *et al.*[25] reported that 26 nonsmoking hospitalized patients in New York City had a mean value of 1.0% carboxyhemoglobin. The carbon monoxide concentrations in the hospital were similar but slightly below those found outdoors at the same time. These last two studies are particularly important because the carboxyhemoglobin was measured by gas chromatography rather than by the spectrophotometric techniques used by the other investigators.

Carboxyhemoglobin concentration in smokers depends on the number of cigarettes smoked, degree of inhalation, and other factors. The carbon monoxide content of cigarette smoke is up to 5% by volume,[180] and about 80% of carbon monoxide produced by smoking cigarettes is retained. Carboxyhemoglobin levels as high as 15% have been reported in chain smokers, but values are usually in the 3–8% range in most smokers.

5

Effects on Man and Animals

UPTAKE OF CARBON MONOXIDE

Carbon monoxide in the body comes from two sources: endogenous, from the breakdown of hemoglobin and other heme-containing pigments; and exogenous, from inhalation. The catabolism of pyrrole rings is the source of the endogenous production of carbon monoxide, which in adults normally leads to carbon monoxide production of about 0.4 ml/hr (STP).[95] This can be increased by hemolytic anemias[96] and the induction of hepatic cytochromes from taking drugs.[88] Dihalomethanes may have increased endogenous carbon monoxide production and markedly elevated carboxyhemoglobin.[18,123,241,385,421]

Inhalation is the first step in the process of exogenous carbon monoxide uptake, followed by an increase in carbon monoxide concentration in the alveolar gas with diffusion from the gas phase through the pulmonary membrane and into the blood. The rate of uptake into the body is limited by the rate of diffusion from the alveoli and the combination with the blood. When the concentration of carbon monoxide is very high, the rate of uptake may be partially limited by the amount that can be inhaled with each breath.

Using modern concepts of the physiologic factors that determine carbon monoxide uptake and elimination, Coburn *et al.*[91] developed an equation for calculating the blood carboxyhemoglobin as a function of time. The basic differential equation was:

$$\frac{d(\mathrm{CO})}{dt} = \dot{V}_{\mathrm{CO}} - \frac{[\mathrm{HbCO}]}{[\mathrm{HbO_2}]} \times \frac{\bar{p}_{\mathrm{C}}\mathrm{O_2}}{M} \times \frac{1}{\dfrac{1}{D_L} + \dfrac{P_B - 47}{\dot{V}_A}} + \frac{P_I\mathrm{CO}}{\dfrac{1}{D_L} + \dfrac{P_B - 47}{\dot{V}_A}}$$

where:

$d(\mathrm{CO})/dt$ is the rate of change of carbon monoxide in the body,
$\dot{V}_{\mathrm{CO}}$ is the carbon monoxide production rate,
[HbCO] is the concentration of carbon monoxide in the blood,
$[\mathrm{HbO_2}]$ is the concentration of oxyhemoglobin,
$\bar{p}_{\mathrm{C}}\mathrm{O_2}$ is the mean pulmonary capillary oxygen pressure,
M is the Haldane constant (220 for pH 7.4),
D_L is the diffusion capacity of the lungs,
P_B is the barometric pressure,
$\dot{V}_A$ is the alveolar ventilation rate, and
$P_I\mathrm{CO}$ is the inspired carbon monoxide pressure.

This equation was designed to investigate the measurement of blood carboxyhemoglobin as an indicator of the rate of carbon monoxide production. Its solution could therefore be based on the assumption that the mean pulmonary capillary oxygen pressure, $\bar{p}_{\mathrm{C}}\mathrm{O_2}$, and the concentration of oxyhemoglobin, $[\mathrm{HbO_2}]$, were constant and independent of the concentration of carboxyhemoglobin, [HbCO]. With these assumptions a solution of the equation became possible. Researchers who have used the Coburn *et al.* solution have accepted these assumptions.

In the general case, however, the oxyhemoglobin concentration depends on the carboxyhemoglobin concentration in a complex way and a solution is only possible using special computer methods. A second approximation solution of the Coburn *et al.* equation permits the evaluation of the kinetics of the washing in and washing out of carbon monoxide over a fairly wide range of carboxyhemoglobin concentrations.* The basic assumptions on which the differential equation and its solutions are developed are described in the original paper by Coburn *et al.*[91] These assumptions are not restrictive, however, and therefore the solutions are applicable generally. The inspired gas is assumed to be ambient air to which carbon monoxide is added.

*We are indebted to Dr. Alan Marcus and Mr. Philip Becker of the University of Maryland at Baltimore County for providing both this solution and the computer printouts.

Besides the carbon monoxide production rate, the principal factors that determine the rate of uptake or release of carbon monoxide from the body are: the concentration of carbon monoxide inspired; the diffusion capacity, D_L, a function of body size and to some extent the level of exercise; the alveolar ventilation, $\dot{V}_A$, also dependent on the amount of exercise; the mean pulmonary capillary oxygen pressure, $\bar{p}_cO_2$, a function both of the barometric pressure and the health of the lungs; and the blood volume, determined by the body size. Thus, the principal factors related to the change in carboxyhemoglobin concentration after exposure are: concentration of carbon monoxide inspired, endogenous carbon monoxide production, amount of exercise, body size, lung health (including diffusion capacity), and barometric pressure.

In addition to exposure from ambient sources, tobacco smoking, particularly of cigarettes, is an important special case of carbon monoxide exposure. The theory predicts that a smoker with a normal concentration of carboxyhemoglobin during the day from ambient carbon monoxide sources will have an additive amount of carboxyhemoglobin from smoking. Recent measurements by Smith (unpublished data) in Calgary, Alberta, of ambient exposures and alveolar carbon monoxide concentrations have shown that the theory predicts end-of-day carboxyhemoglobin with reasonable accuracy as long as the ambient carbon monoxide does not fluctuate rapidly during the day, as it does with certain occupational exposures.

As illustrated in Figures 5-1–5-4, the theory can also be applied to the influence of various factors, such as cigarette smoking, lung health, duration of exposure, and altitude, on the rate of change of carboxyhemoglobin during the daily activity cycle and to its concentration at night during sleep.

In the following graphs the variables used in the Coburn *et al.* equation are for adult subjects at sea level. Any modifications are indicated in the legends to the figures.

The values assumed for the general case are:

Haldane constant = 220 (for pH 7.4)
$\bar{p}_cO_2$ (mean pulmonary capillary oxygen, partial pressure) = 95 torr
D_L (diffusion capacity) = 20 $ml \cdot min^{-1} torr^{-1}$
Sleeping $\dot{V}_A$ (alveolar ventilation) = 3 $liters \cdot min^{-1}$
Light work $\dot{V}_A$ (light exercise) = 5 $liters \cdot min^{-1}$
V_b (blood volume) = 5 liters

The peak values reached are less than the equilibrium values (Table 4-6). For 10 ppm the peak percent of carboxyhemoglobin reached on the

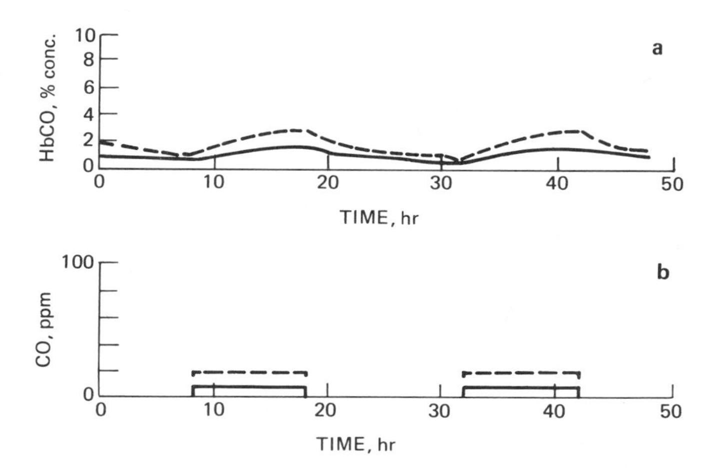

FIGURE 5-1 (a) Carboxyhemoglobin in a nonsmoker exposed to carbon monoxide during the period 8 a.m. to 6 p.m. and not exposed at other times; sleep from 10 p.m. to 6 a.m.; light exercise when awake. Zero time is midnight on the first day. The initial carboxyhemoglobin is an arbitrarily chosen value. Solid line: ambient carbon monoxide 10 ppm during day; dashed line: ambient carbon monoxide 20 ppm during day. (b) Carbon monoxide exposure pattern used to calculate above carboxyhemoglobin.

second day (when the initial conditions no longer are important) is 1.5% compared to 1.85% predicted for equilibrium (Figure 5-1). At 20 ppm the peak value reached at the second day is 2.86% compared to 3.29% for equilibrium. These results show that, if exposure to carbon monoxide occurs only during the working day from ambient or occupational sources, the carboxyhemoglobin reached will be well below equilibrium for persons not doing heavy work.

By the second day, a consistent daily pattern of carboxyhemoglobin has been achieved. The peak carboxyhemoglobin reached in the evening in this light inhaler is 4.5%. The carboxyhemoglobin remains well above that from endogenous production even when smoking is stopped during sleep (Figure 5-2).

A comparison of the same computer run made without the ambient exposure of 10 ppm but with the same smoking parameters confirms that the ambient exposure is almost entirely additive to the cigarette exposure (Figure 5-3).

The peak carboxyhemoglobin reached on the second day is 5.68% for a normal subject and 7.11% for a patient with chronic lung disease (Figure

5-4). The difference is partly due to the reduced rate of carbon monoxide loss during sleep but is mainly due to the low value of $\bar{p}_{C}O_2$ applicable to the patient with chronic lung disease.

PHYSIOLOGICAL EFFECTS

General

The respiratory and cardiovascular systems working together transport oxygen from the ambient air to the various tissues of the body at a rate sufficient to maintain tissue metabolism. In one step in this overall process, oxygen is carried by the blood from the lungs to extrapulmonary tissues. Nearly all the oxygen in the blood is reversibly bound to the hemoglobin contained in the red blood cells. The most important chemical characteristic of carbon monoxide is that it, too, is reversibly bound by hemoglobin, competing with oxygen for the binding sites on the hemoglobin molecule. Hemoglobin's affinity for carbon monoxide is more than 200 times greater than for oxygen. Therefore, carbon monoxide

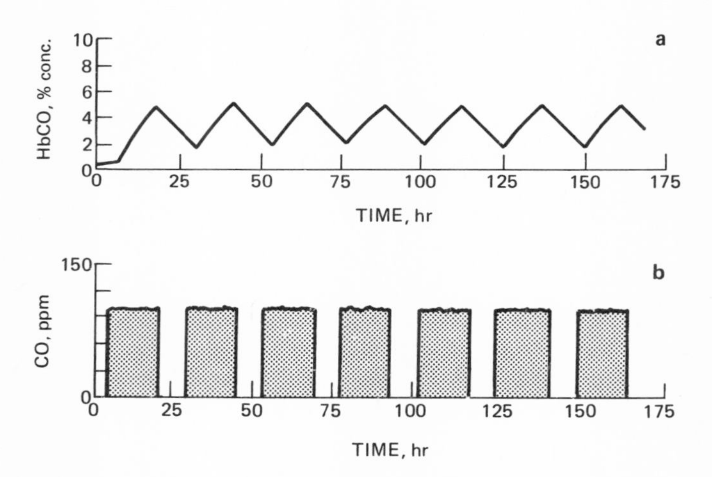

FIGURE 5-2 (a) Carboxyhemoglobin in a smoker during 1 week, smoking 48 cigarettes per day, but with relatively light inhalation. Sleep and awake conditions as in Figure 5-1. At zero time (midnight), the initial carboxyhemoglobin is assumed to be 0.4%, which is consistent with endogenous production only. (b) The pattern of exposure to carbon monoxide applicable to the upper graph. Each single vertical line represents a cigarette smoked for 5 min, with a mean alveolar pCO of 100 ppm during that 5-min interval (light inhalation).

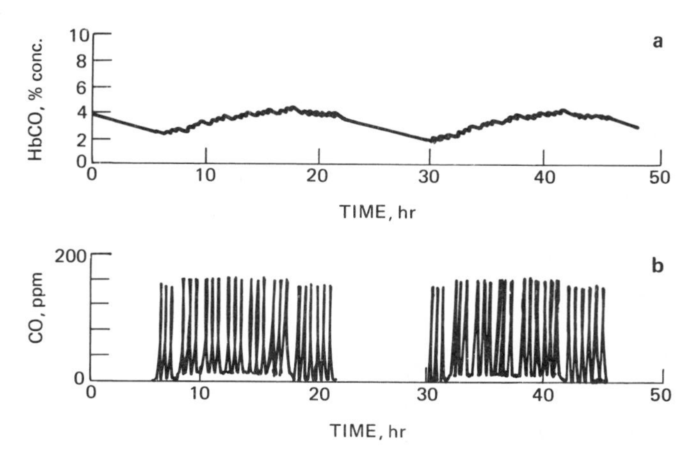

FIGURE 5-3 (a) Carboxyhemoglobin in a smoker exposed to 10 ppm during the working day. Twenty-four cigarettes smoked per day. The peak carboxyhemoglobin reached on the second day is 4.35%. (b) Exposure pattern to carbon monoxide applicable to (a). Each spike represents smoking one cigarette for 5 min with a mean alveolar pCO of 150 ppm (moderate inhalation),

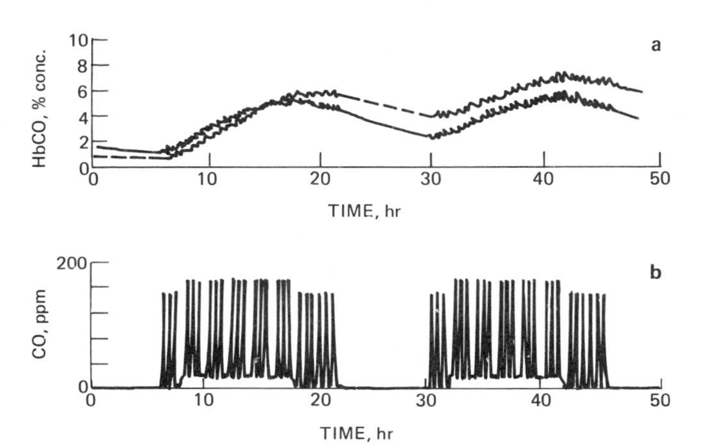

FIGURE 5-4 (a) Carboxyhemoglobin in a smoker living in Denver, Colorado (5,200-ft altitude) and exposed during the working day (8 a.m.-6 p.m.) to carbon monoxide at 20 ppm, a concentration sometimes found in work environments. The smoking conditions were the same as in Figure 5-3. Solid line: normal lungs (D_L = 20 ml·min^{-1} torr^{-1}; $\bar{p}_{C}O_2$ = 75 torr). Dashed line: chronic lung disease (D_L = 10 ml·min^{-1} torr^{-1}; $\bar{p}_{C}O_2$ = 50 torr).

can seriously impair the transport of oxygen even when present at very low partial pressures.

The proportion of hemoglobin combined with carbon monoxide at any time is determined not only by the partial pressure of carbon monoxide but also by that of oxygen. The approximate relation stated by Haldane and his associates,[127] in 1912, showed that the ratio of the concentrations of carboxyhemoglobin (HbCO) and oxyhemoglobin (HbO_2) is proportional to the ratio of the partial pressures of carbon monoxide and oxygen:

$$\frac{[\mathrm{HbCO}]}{[\mathrm{HbO_2}]} = M \frac{(\mathrm{pCO})}{(\mathrm{pO_2})}$$

The constant M is about 210 for human blood. The accuracy of this expression has been questioned, particularly when a large fraction of the hemoglobin present is not combined either with oxygen or with carbon monoxide.[213] The equation approximates the actual relationship closely, however, and has proven very useful for quantitatively analyzing the influence of carbon monoxide on oxygen transport by the blood.

The oxyhemoglobin dissociation curve (Figure 5-5) describes oxygen transport. Under normal conditions when the arterial blood has been equilibrated in the lungs with an oxygen partial pressure (pO_2) of 90–100 mm Hg, it contains slightly less than 20 vol% of oxygen bound to hemoglobin (point a). As the blood flows through the various tissues of the body, oxygen diffuses out of the capillaries to meet metabolic needs. The pO_2 falls as the oxygen content of the blood is reduced by an amount determined by the metabolic rate of the tissue and the capillary blood flow. For a typical tissue or for the entire body, the venous oxygen content is about 5 vol% lower than the arterial, and the venous pO_2 is about 40 mm Hg (point v).

The normal oxygen transport by the blood may be impaired by a variety of factors. In Figure 5-5 the effects of carbon monoxide on venous pO_2 are shown and compared to the normal state and to the effects of a decrease in blood hemoglobin content (anemia). Arterial pO_2 and oxygen content are indicated by the symbols a and a′ and the symbols v, v_1', and v_2' indicate venous pO_2 and oxygen content assuming the arteriovenous (A–V) oxygen content difference remains constant at 5 ml/100 ml. This figure shows that there are two consequences of increasing carboxyhemoglobin, an effect like anemia that results in decrease of pO_2 in tissue capillary blood and a shift to the left of the oxyhemoglobin dissociation curve, which further decreases the pO_2 in the capillary blood in peripheral tissues. For the sake of clarity, an example of severe carbon

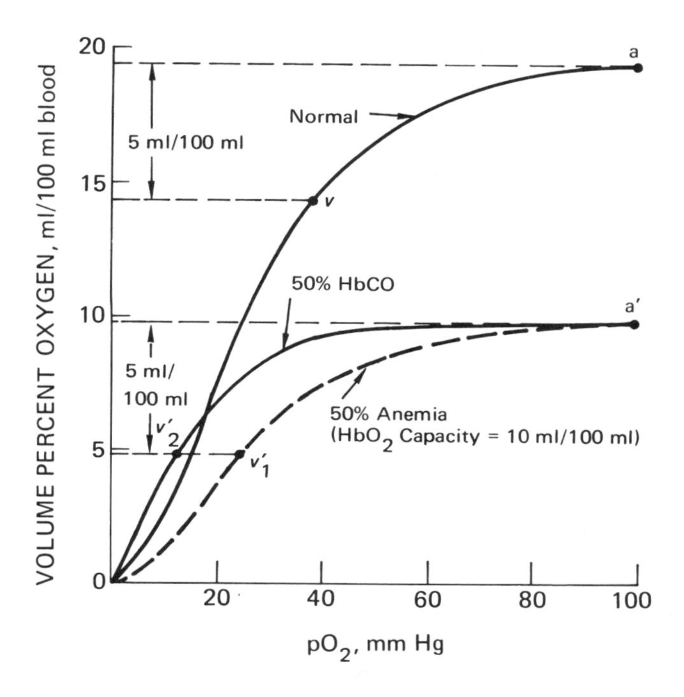

FIGURE 5-5 Oxyhemoglobin dissociation curves of normal human blood, of blood containing 50% carboxyhemoglobin, and of blood with a 50% normal hemoglobin concentration due to anemia. (From Rahn and Fenn[362] and Roughton and Darling.[387])

monoxide poisoning has been used in this discussion. Under conditions of mild carbon monoxide poisoning, as shown in Figure 5-6, the effects are qualitatively similar but less severe.

The ultimate indicator of the oxygen transport system's success is whether the various tissues of the body receive oxygen at a rate adequate to sustain their normal function. Aerobic metabolic processes depend on the maintenance of tissue pO_2 above some critical point that varies among different tissues. Intracellular gas pressures are difficult to measure directly, but, other things being equal, changes in capillary pO_2 reflect changes in tissue pO_2. In the absence of arteriovenous shunts, the pO_2 of the venous blood draining a tissue is equal to the pO_2 at the venous end of its capillaries. Therefore, the pO_2 of venous blood roughly and indirectly indicates the adequacy of tissue oxygenation.

The venous pO_2 values expected to result from various degrees of carbon monoxide poisoning can be easily calculated.[30,144,347] As shown in Figure 5-7, if the blood flow and metabolic rate remain constant, equilibration with carbon monoxide at 200 ppm will lower venous pO_2 from about 40 mm Hg to less than 30 mm Hg. A similar degree of venous hypoxemia results from a 35% decrease in the oxygen capacity due to anemia or from ascent to an altitude of 12,000 ft (3,660 m). The broken lines (Figure 5-7) show the effects of changes in blood flow or metabolic rate. The upper curve indicates, for example, that the effect of mild carbon monoxide poisoning on venous pO_2 can be offset by a modest increase in blood flow. The lower curve indicates that individuals or tissues with high rates of oxygen consumption or low rates of blood flow may be particularly susceptible to carbon monoxide poisoning.

This analysis assumes that alveolar ventilation, metabolic rate, and blood flow all remain constant and that little or no mixed venous blood is shunted through the lungs without being equilibrated with alveolar air. It is well known, however, that 1–2% of the cardiac output is shunted

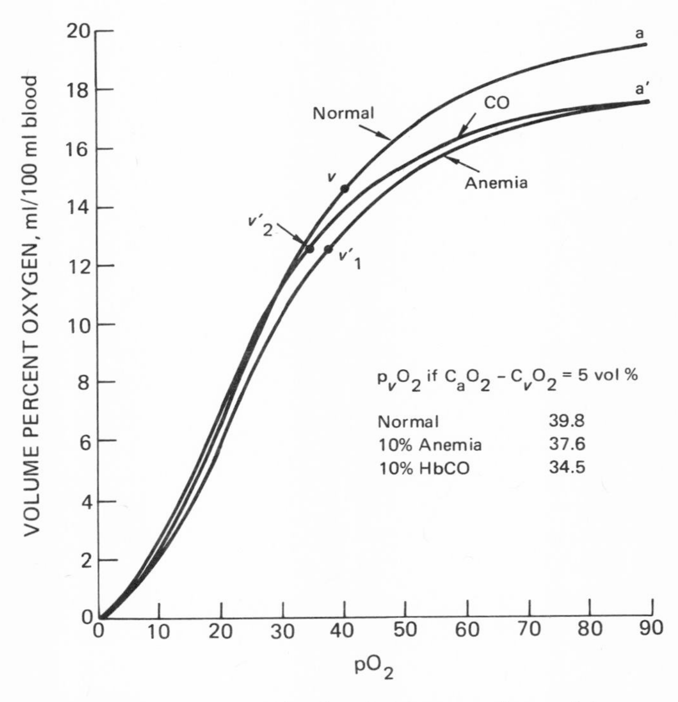

FIGURE 5-6 Oxyhemoglobin dissociation curve of normal human blood containing 10% carboxyhemoglobin and blood with 10% decrease in hemoglobin concentration (10% anemia). (Symbols v, v_1', and v_2' have same meaning as in Figure 5-5.)

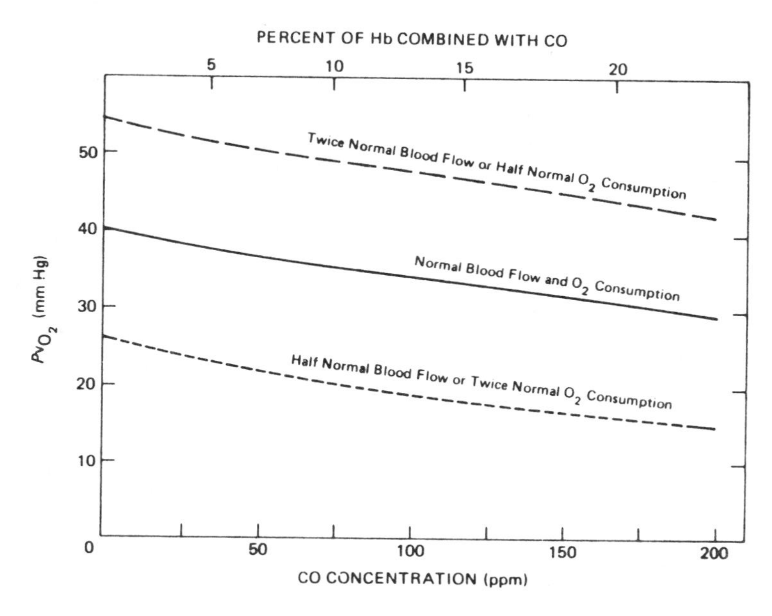

FIGURE 5-7 Effect of carbon monoxide on venous pO_2. The arteriovenous oxygen content difference with normal blood flow and oxygen consumption is assumed to be 5 vol%. (Modified slightly from Figure 3 of Bartlett.[30])

past the alveolar capillaries in normal subjects and that much larger right-to-left shunts exist under disease conditions. Mixed venous blood flowing through these right-to-left shunts combines with blood that has undergone gas exchange in the pulmonary capillaries to form the mixed arterial blood. Because the oxygen content of the shunted blood is lower than that of the end-capillary blood with which it mixes, the resulting oxygen content of the arterial blood is lower than that of the end-capillary blood, and arterial pO_2 is somewhat lower than mean alveolar pO_2. Brody and Coburn[54] have pointed out that if the oxygen content of the mixed venous blood is abnormally low, as in anemia or carbon monoxide poisoning, the effect of the shunted blood in lowering the arterial pO_2 will be greater than normal, resulting in a small increase in the alveolar-arterial oxygen pressure difference (A-a DO_2). If the mixed venous pO_2 and the right-to-left shunt remain constant, the change in the shape of the oxyhemoglobin curve due to the presence of carbon monoxide also increases the A-a DO_2. A similar phenomenon occurs when some lung regions have nonuniform ventilation perfusion ratios, the case in many types of cardiopulmonary disease. Figure 5-8, taken from the work of

Brody and Coburn,[54] shows that slight increases in carboxyhemoglobin concentration have little or no influence on the alveolar-arterial oxygen pressure difference (A–a DO_2) in normal subjects, but increase the A–a DO_2 in patients with large intracardiac right-to-left shunts or with chronic lung disease and mismatching of ventilation and perfusion of carbon monoxide. This phenomenon adds a component of arterial hypoxia to the effect of carbon monoxide on oxygen transport in such patients.

It has been suggested[57] that cytochrome P-450 is important in oxygen transport in cells and that one effect of increased carbon monoxide

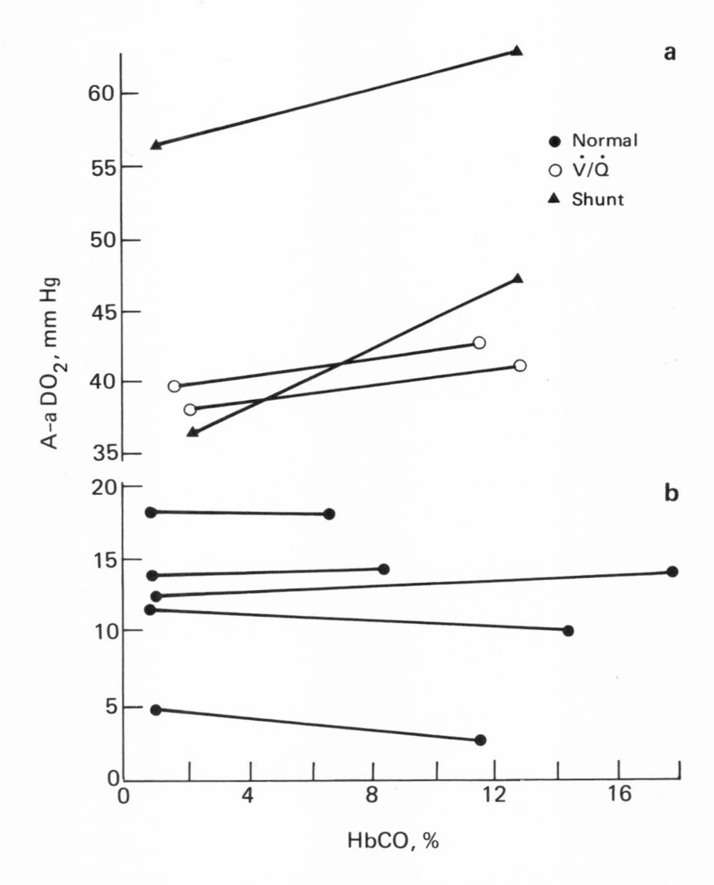

FIGURE 5-8 Effect of carbon monoxide administration on the alveolar arterial pO_2 difference in normal subjects (●), in patients with $\dot{V}_A/\dot{Q}$ abnormalities due to chronic lung disease (○), and in patients with intracardiac right-to-left shunts (▲). (Reprinted with permission from Brody and Coburn.[54])

tension is to block transport facilitated via this mechanism. At present, however, there is no convincing evidence to support this postulate.

Intracellular Effects of Carbon Monoxide

TISSUE CARBON MONOXIDE TENSION

Intracellular effects of carbon monoxide depend on the carbon monoxide partial pressures (pCO) in the tissues. Calculations have been made of tissue pCO from blood carboxyhemoglobin and an assumed mean capillary oxygen partial pressure pO_2, using the Haldane equation. At a blood carboxyhemoglobin concentration of 5%, pCO is 2×10^{-2} mm Hg (assuming the mean capillary pO_2 is 40–50 mm Hg), about five times greater than when the carboxyhemoglobin is normal. This is about 60–70% of the inspired pCO for a steady-state condition.

Göthert and co-workers[168] have recently (1970) estimated tissue pCO in the rabbit peritoneum from measurements of pCO in an air pocket in the peritoneal cavity. Measurements were made with rats, guinea pigs, or rabbits breathing carbon monoxide at 86–1,000 ppm. The pCO in the gas pocket was 42–69% of the partial carbon monoxide pressure in the alveolar air. This figure probably underestimates tissue pCO in tissues where the ratio of oxygen extraction to blood flow is less than in peritoneum. Campbell[68] made similar measurements of pCO in a gas bubble in the peritoneum of mice.

INTRACELLULAR pO_2

The pO_2 in proximity to intracellular compounds that bind carbon monoxide is a critical factor in possible intracellular effects of carbon monoxide,[89] since oxygen and carbon monoxide binding are competitive. In general, recent research is pointing to the presence of a lower intracellular pO_2 than previously thought. Studies using polarographic microelectrodes in liver (Kessler *et al.*[224]) demonstrated a tissue pO_2 of <10 mm Hg in 10–20% of their penetrations. Whalen[484] found that intracellular pO_2 in skeletal and cardiac muscle averaged 5-6 mm Hg. Computing mean myoglobin pO_2 from carbon monoxide binding to myoglobin also gives a normal pO_2 value of 4–7 mm Hg (Coburn *et al.*[94]). Tissue pO_2 levels in brain are somewhat higher. Since we have no idea about either pO_2 gradients or compartmentalization in cells, it is possible that intracellular pO_2 at the site of carbon monoxide binding compounds are considerably below these levels.

Chance *et al.*[74] indicate that the pO_2 in mitochondria under conditions of tight respiratory coupling and the presence of ADP (state 3) is inhibited 50% at 0.01–0.05 mm Hg. Since it is possible that the pO_2 in mitochondrial cristae is as low as 0.01 mm Hg, even though mean cytoplasmic pO_2 is 4–6, later in this chapter we use this low value in calculations of possible effects of carbon monoxide on cytochrome a_3 function.

Zorn[507] studied the effects of carbon monoxide inhalation on brain and liver pO_2, using Lubbers' platinum electrode and surface electrodes. It was found in both tissues that tissue pO_2 fell, even at carboxyhemoglobin of 2% saturation, and that the fall was almost directly related to the increase in carboxyhemoglobin. For a 1% fall in oxyhemoglobin saturation due to an increase in carboxyhemoglobin, pO_2 decreased 0.2 to 1.8 mm Hg. This is a particularly nice approach since, if carbon monoxide had only an intracellular effect, tissue pO_2 would be expected to increase. When the experimental animal breathed air not containing carbon monoxide, tissue pO_2 returned toward normal.

Weiss and Cohen[476] have reported similar effects of breathing 80 and 160 ppm carbon monoxide for 20 min (resulting in carboxyhemoglobin saturation of less than 3.3%) on rat brain cortex pO_2 and rat biceps brachii muscle, as measured with a bare platinum electrode.

DIRECT CARBON MONOXIDE EFFECTS ON INTRACELLULAR PROCESSES—CYTOCHROME a_3 AND MITOCHONDRIAL ELECTRON CHAIN TRANSPORT

Carbon monoxide affinity to intracellular compounds has customarily been given in terms of the Warburg partition coefficient:

$$K = [n/1 - n]\,[CO/O_2]$$

where n is the fraction bound to carbon monoxide and CO/O_2 is the ratio of carbon monoxide to O_2. The data are usually given where $n = 0.5$, which gives the ratio of carbon monoxide to oxygen for 50% saturation with carbon monoxide. There is apparently no new information about K for a_3, which has been widely quoted to be 2.2–28.[494] In the normal aerobic steady state in the rapidly respiring state 3, the concentration of reduced cytochrome a_3 is very low, probably less than 0.1%. Because only reduced cytochrome a_3 binds carbon monoxide, the affinity of carbon monoxide for cytochrome a_3 under this condition is small. In coupled mitochondria at low ADP concentration (state 4), the steady-state concentration of reduced cytochrome a_3 may be even lower.

Earlier studies of Chance[73] have not been appreciated in previous publications concerning possible effects of carbon monoxide on mitochondria. He studied the transient from anoxia to normoxia in pigeon heart mitochondria in both the absence and presence of carbon monoxide. It was found in uncoupled mitochondria that CO/O_2 ratios of 0.2 caused a marked delay in this transient. Thus these mitochondria were markedly more sensitive to the effects of carbon monoxide when studied in this state. Chance also points out that the dissociation of carbon monoxide from reduced a_3 is so slow that it should take 3-4 min for one-half unloading to occur; thus, after a hypoxic episode, cytochrome a_3 function can be influenced by much lower tissue pCO.

Recent data have suggested that mitochondrial respiration may be more sensitive to carbon monoxide under some conditions than previously indicated; however, there is no solid evidence for implicating this system at low blood carboxyhemoglobin levels. Even using CO/O_2 ratios of 0.2 for 50% binding to a_3, which Chance found in uncoupled mitochondria during transients, and a pO_2 of 0.01 mm Hg, computed tissue pCO at 5% carboxyhemoglobin of 2×10^{-2} mm Hg is slightly below that necessary to cause 50% binding to a_3; however, at a carboxyhemoglobin of 10%, tissue pCO *is* high enough. There is a probability that binding of carbon monoxide to a_3 is physiologically significant during tissue hypoxia at very low blood carboxyhemoglobin levels. Previous reviews of possible effects of carbon monoxide on a_3 have failed to point out that the chemical constants and investigations are never performed at 37 C but at 25 C or less.

A recent line of investigation[342] has revealed that heart mitochondria isolated after chronic arterial hypoxemia have a higher state 3 mitochondrial oxygen uptake, per gram of protein, than mitochondria isolated from an animal which was not chronically hypoxemic. These data may be pertinent to acclimatization to carbon monoxide, which has been demonstrated in experimental animals.

CYTOCHROME P-450

The Warburg coefficient for the cytochrome P-450 has been quoted by Estabrook *et al.*[137] to be 1-5. Calculations similar to those performed for cytochrome a_3 in the preceding section suggest, using values of 5 to 10 mm Hg for microsomal pO_2, that tissue pCO is 1 to 2 orders of magnitude too low for carboxyhemoglobin at less than 15% saturation to have an effect on the cytochrome P-450 system.

Recent advances in our knowledge of the effects of carbon monoxide on the cytochrome P-450 system are discussed below. Estabrook *et al.*[137] found that, under conditions of rapid electron transport through the cytochrome P-450 system, sensitivity to carbon monoxide increased.

In the presence of high concentrations of reducing equivalents and substrate, the CO/O_2 ratio necessary for 50% binding was as low as 0.2, whereas with slow electron transport the system becomes almost completely refractory to carbon monoxide. Since carbon monoxide sensitivity does vary under changing conditions, it is possible that in some conditions carbon monoxide sensitivity might increase to levels where the cytochrome P-450 system is influenced by carbon monoxide tension in tissues at low carboxyhemoglobin.

Data are now available on the effects of elevated carboxyhemoglobin on the cytochrome P-450 system. Rondia[381] found that in rats exposed to 60 ppm there was a decreased ability of liver to metabolize 3-hydroxybenzo[a]pyrene. Montgomery and Rubin[318,319] found prolonged sleeping time in the presence of 20% carboxyhemoglobin in rats given hexobarbital. However, when they compared these effects owing to carbon monoxide poisoning to those of hypoxic hypoxia, looking at the effects with an equivalent arterial oxyhemoglobin percent saturation, it was discovered the effects on sleeping time were greater owing to hypoxic hypoxia than to carbon monoxide poisoning. In subsequent studies, Roth and Rubin[382-384] have found that the greater effect with hypoxic hypoxia was due to a decrease in hepatic blood flow. The effect of carbon monoxide is probably due to a fall in capillary pO_2 rather than carbon monoxide binding to cytochrome P-450. Coburn and Kane[92] found that 10-12% carboxyhemoglobin concentration in anesthetized dogs resulted in inhibition of loss of hemoglobin-haptoglobin from the plasma. It is now known that catabolism of hemoglobin-haptoglobin is mediated by a cytochrome P-450-linked system, but this could be due to a blood flow effect. Rikans[371] has reported evidence for a carbon monoxide binding compound in solubilized rat hepatic microsomes that is distinct from cytochrome P-450; however, the chemistry and relative affinities for carbon monoxide and oxygen, as well as its function, have not been delineated.

MYOGLOBIN

The Warburg partition coefficient for the reaction of carbon monoxide with oxymyoglobin is 0.04. Myoglobin is probably the intracellular hemoprotein most likely to be involved in toxic effects of carbon monoxide.

Coburn *et al.*[94] have measured the ratio MbCO/HbCO and found it to be approximately 1 and constant, even with increases in blood carboxyhemoglobin exceeding 20% saturation. Thus, at 5% concentration of carboxyhemoglobin, 5% of the myoglobin should be bound to carbon monoxide. It is difficult to interpret this in terms of toxic effects of carbon monoxide on skeletal muscle, smooth muscle, or heart muscle since the function of myoglobin is not clearly defined, nor is it known where it is located in the cell. In the heart, there is evidence that myoglobin buffers changes in oxygen tension in close proximity to mitochondria during muscle contraction. Another possible function of these compounds is to facilitate oxygen transport across the cytoplasm from cell membrane to mitochondrion. Like other intracellular hemoproteins, carbon monoxide binding is markedly increased in the presence of tissue hypoxia.

There is evidence (by Clark and Coburn[84]) of significant carbon monoxide shifts out of blood, presumably into skeletal muscle, during short-term bicycle exercise at maximal rate of oxygen uptake.

EFFECTS OF CARBON MONOXIDE ON THE PREGNANT WOMAN, DEVELOPING EMBRYO, FETUS, AND NEWBORN INFANT

Insufficient knowledge exists about the biological effects of carbon monoxide during intrauterine development and the newborn period. Several studies report decreased birthweights and increased mortality in the progeny of animals exposed to relatively high carbon monoxide concentrations, but few studies have reported on the more subtle effects at lower concentrations. Most of the evidence supporting carbon monoxide effects on the fetus is inferred from data on maternal smoking rather than from studies of its effects per se. This section reviews both what is known and what is not known about carbon monoxide effects on the developing embryo, fetus, and newborn infant; carbon monoxide exchange between the mother and the fetus under both steady-state and non-steady-state conditions; and the mechanisms by which it interferes with oxygenation of the fetus.

Interrelations of Carboxyhemoglobin Concentrations in the Mother and Fetus

MATERNAL CARBOXYHEMOGLOBIN LEVELS

The carboxyhemoglobin concentration in the blood of normal nonsmoking pregnant women varies from 0.5 to 1%.[22,275] In addition to those factors affecting carboxyhemoglobin in the nonpregnant person,[91] maternal carboxyhemoglobin concentration, [$HbCO_m$], reflects the endogenous carbon monoxide production by the fetus and its rate of exchange across the placenta.[279] Fetal endogenous carbon monoxide production accounts for about 3% of the total carboxyhemoglobin present in the blood of a normal pregnant woman.

FETAL CARBOXYHEMOGLOBIN

Under steady-state conditions, the concentration of human fetal carboxyhemoglobin, [$HbCO_f$], is greater than that in maternal blood (see Figure 5-9). The wide disparity in the reported values of both human fetal carboxyhemoglobin concentrations and the ratio of fetal to maternal carboxyhemoglobin concentrations, [$HbCO_f$]/[$HbCO_m$],[275] probably results from a number of factors. These include collecting the samples under non-steady-state conditions and using different methods to analyze for carbon monoxide. While the fetal carboxyhemoglobin concentration varies as a function of the concentration in the maternal blood, it also depends upon the rate of fetal carbon monoxide production, placental carbon monoxide diffusing capacity, the relative affinity of both fetal and maternal hemoglobin for carbon monoxide as compared to its affinity for oxygen, and the relative affinity of blood for these two gases.

The ratio of the fetal to maternal carboxyhemoglobin concentration during the steady state depends on several factors. Assuming that the

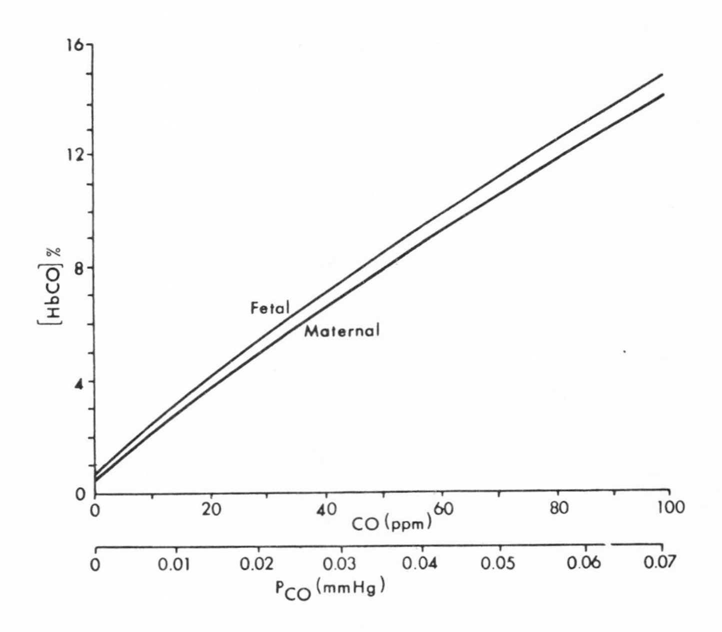

FIGURE 5-9 Relation of human maternal and fetal carboxyhemoglobin concentrations under steady-state conditions as a function of both carbon monoxide partial pressure (mm Hg) and inspired air concentrations in parts per million. (Reprinted with permission from Hill *et al.*[195])

carbon monoxide partial pressure in maternal blood, pCO_m, equals the carbon monoxide partial pressure in fetal blood, pCO_f, the Haldane equation $pCO = ([HbCO] \times pO_2)/([HbO_2] \times M)$ may be equated for maternal and fetal blood. The result when rearranged becomes:

$$\frac{[HbCO_f]}{[HbCO_m]} = \frac{[HbO_{2f}]\, pO_{2m}\, M_f}{pO_{2f}\, [HbO_{2m}]\, M_m} \quad (1)$$

where the ratios of oxyhemoglobin concentration to the oxygen partial pressure for both fetal and maternal blood equal oxygen affinities of fetal and maternal blood determined from the oxyhemoglobin saturation curves at the mean oxygen tension of blood in the placental exchange vessels; and M_f and M_m are the relative affinities of fetal and maternal blood, respectively, for carbon monoxide as compared to oxygen. Thus, the ratio of the concentrations of fetal to maternal carboxyhemoglobin, $[HbCO_f]/[HbCO_m]$, during the steady state depends on both the relative affinities of fetal and maternal hemoglobin for oxygen and the ratio of the relative affinity of fetal and maternal blood for carbon monoxide and oxygen, M_f/M_m.

TIME COURSE OF CHANGES IN FETAL AND MATERNAL CARBOXYHEMOGLOBIN

The relation of carboxyhemoglobin concentrations to inspired carbon monoxide concentrations in adult humans has been experimentally determined in several studies. Recently, Longo and Hill[278] studied these relations in pregnant sheep with catheters chronically implanted in both maternal and fetal blood vessels. They exposed ewes to inspired carbon monoxide concentrations of up to 300 ppm. At 30 ppm, maternal carboxyhemoglobin concentrations increased over a period of 8 to 10 hr, equilibrating at about 4.6 ± 0.3 (SEM)%. The fetal carboxyhemoglobin concentration increased more slowly, equilibrating in 36 to 48 hr at about 7.4 ± 0.5% (Figure 5-10). At 50 ppm, the time courses were similar, with maternal and fetal steady-state values of 7.2 ± 0.5% and 11.3 ± 1.1%, respectively. At 100 ppm, maternal and fetal steady-state values were 12.2 ± 1.2% and 19.8 ± 1.4%, respectively. For all three concentrations, the half-times for carbon monoxide uptake by maternal and fetal blood were about 2 and 6 hr, respectively.

For ethical and technical reasons, such experiments cannot be carried out with humans. This same group[195] examined the experimental data and theorized a mathematical model of the interrelations of human fetal and maternal carboxyhemoglobin concentrations. The predicted changes in the carboxyhemoglobin concentrations as a function of the time of

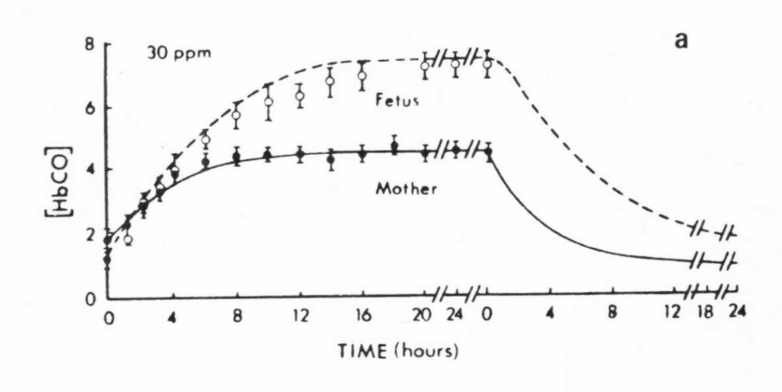
a
30 ppm
Fetus
Mother
[HbCO]
0
2
4
6
8
0 4 8 12 16 20 24 0 4 8 12 18 24
TIME (hours)

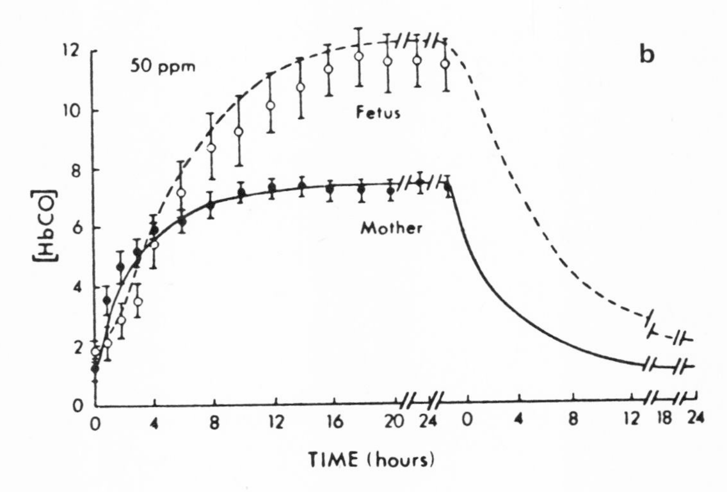
b
50 ppm
Fetus
Mother
[HbCO]
0
2
4
6
8
10
12
0 4 8 12 16 20 24 0 4 8 12 18 24
TIME (hours)

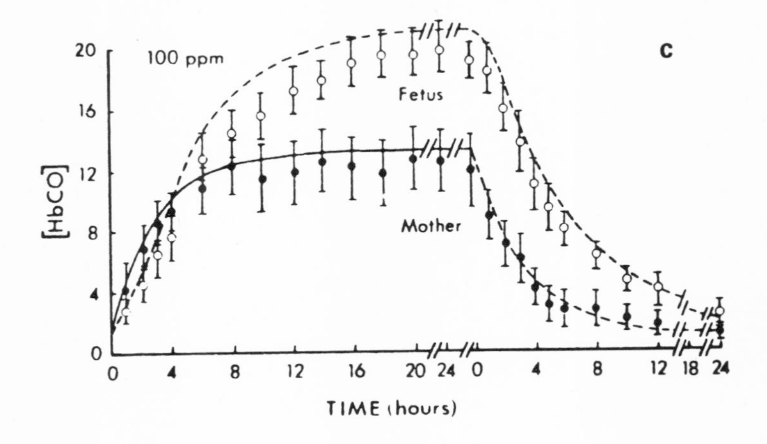
c
100 ppm
Fetus
Mother
[HbCO]
0
4
8
12
16
20
0 4 8 12 16 20 24 0 4 8 12 18 24
TIME (hours)

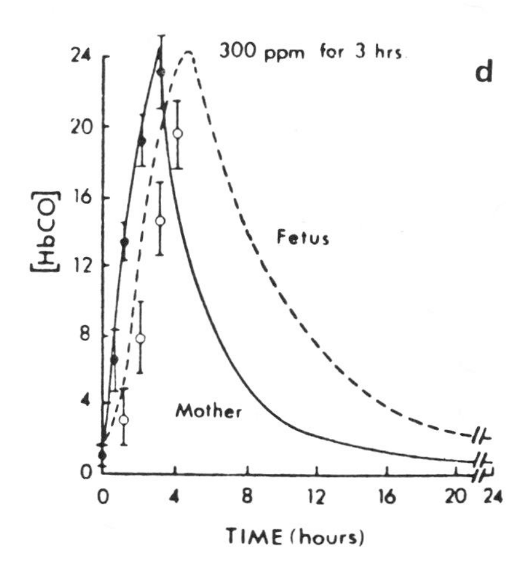

FIGURE 5-10 Time course of carbon monoxide uptake in maternal and fetal sheep exposed to varying carbon monoxide concentrations. The experimental results for the ewe (●) and fetal lamb (○) are the mean values (± SEM) of 9 to 11 studies at each inspired carbon monoxide concentration, except in the case of 300 ppm. Only three studies were performed at that concentration. The theoretical predictions of the changes in maternal and fetal carboxyhemoglobin concentrations for the ewe and lamb are shown by the solid and interrupted lines, respectively. (Reprinted with permission from Longo and Hill.[278])

exposure to inspired carbon monoxide concentrations ranging from 30 to 300 ppm are shown in Figure 5-11.

Several points of interest are: The equilibration for fetal carboxyhemoglobin was not achieved for about 30 to 36 hr; the half-time of the increase in fetal carboxyhemoglobin concentration was about 7.5 hr for all concentrations; the time for fetal carboxyhemoglobin to equal the maternal value varied from 12 to 15 hr; and finally, under steady-state conditions, fetal carboxyhemoglobin concentration was about 15% greater than the maternal. The theoretical relations in humans and the experimental data in sheep are therefore in reasonably good agreement.

THEORETICAL PREDICTION OF FETAL CARBOXYHEMOGLOBIN CONCENTRATION DURING INTERMITTENT MATERNAL CARBON MONOXIDE EXPOSURE

Mothers who smoke cigarettes or who are exposed to excessive amounts of carbon monoxide in the air are subjected to fluctuating concentrations of inspired carbon monoxide. The previously described mathematical model[195] was used to simulate such conditions and to predict the changes in fetal and maternal carboxyhemoglobin concentrations during exposure to various carbon monoxide concentrations for different durations. The time course of carboxyhemoglobin concentrations anticipated if the mother breathed 50 ppm carbon monoxide for a 16-hr period or smoked one and one-half packs of cigarettes a day is shown in Figure 5-12. The peak fetal carboxyhemoglobin concentrations were greater than the maternal concentrations, and the mean carboxyhemoglobin concentrations were 6 and 5.4%, respectively. Similar patterns would be produced by exposure of the mother (and fetus) to other carbon monoxide concentrations with the values reflecting the concentrations. While peak fetal carboxyhemoglobin concentrations were only about 10% greater than maternal, the mean values were about 20% greater. (This difference ranged from 25% greater at 5 ppm to 15% greater at 50 ppm.) The implications for the fetus of this greater carbon monoxide exposure are unknown.

This mathematical model also has been used to predict fetal and maternal carboxyhemoglobin changes following more complicated exposure patterns. Data on carbon monoxide concentrations obtained from the Los Angeles Air Pollution Control District were measured at numerous sites in that city. Figure 5-13 shows the data for a typical site in southern Los Angeles (site number 76 in Torrance) during January 22, 23, and 24, 1974, a Tuesday, Wednesday, and Thursday. The inspired carbon monoxide concentration fluctuated between 0 and 48 ppm (upper curve). The values plotted represent hourly averages. The lower curve shows the calculated maternal and fetal carboxyhemoglobin concentrations, assuming a pregnant woman breathed this air with no additional source of carbon monoxide such as cigarette smoke.[195] Following the peaks of inspired carbon monoxide, the fetal carboxyhemoglobin concentration averaged 3%, while the maternal concentration averaged 2.6%.

These relatively low carboxyhemoglobin concentrations may seem too small to be of much significance. However, several investigators (reviewed in other sections of this report) have demonstrated significant reductions in a number of physiologic functions with blood carboxyhemoglobin levels in the range of 4 to 5%. For the case of a pregnant mother who smoked one to two packs of cigarettes per day, exposure to these elevated ambient carbon monoxide concentrations would be nearly additive. Thus, it can be calculated that the fetal carboxyhemoglobin concentration would be 6 to 7% in the pregnant mother who smoked one pack of cigarettes per day and was exposed to this level of air pollution. Similarly,

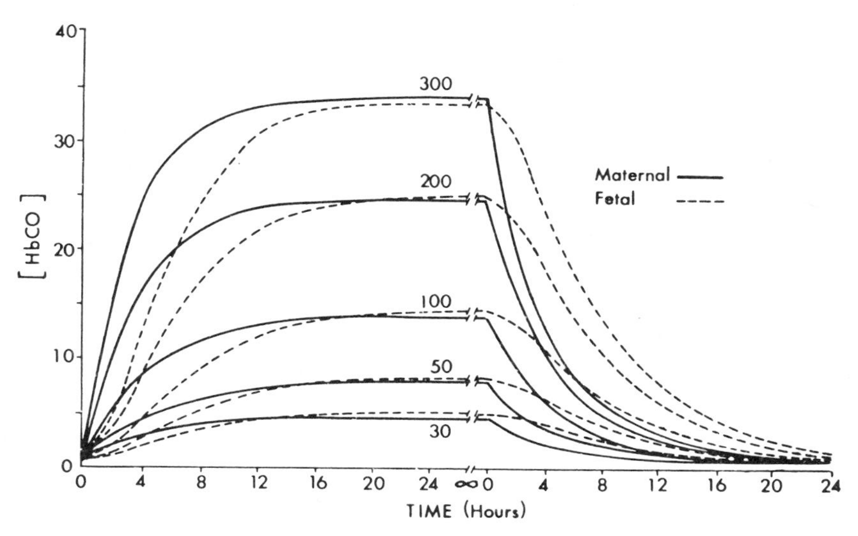

FIGURE 5-11 The predicted time course of human maternal and fetal carboxyhemoglobin concentrations during prolonged exposures to 30, 50, 100, 200, and 300 ppm inspired carbon monoxide concentrations, followed by a washout period when no carbon monoxide is inspired. Note that the fetal carboxyhemoglobin concentrations lag behind those of the mother, but eventually reach higher values in most cases. (Reprinted with permission from Hill *et al.* [195])

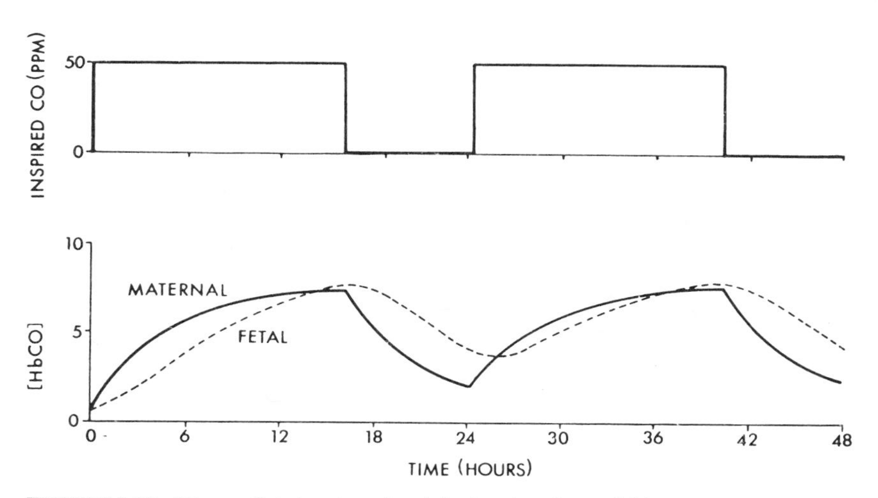

FIGURE 5-12 The predicted maternal and fetal carboxyhemoglobin concentrations when a mother breathes 50 ppm carbon monoxide for a 16-hr period followed by 8 hr during which no carbon monoxide is breathed. This level of carbon monoxide exposure is equivalent to smoking about 1 to 1.5 packs of cigarettes per day, followed by an 8-hr sleep period. (Reprinted with permission from Hill *et al.* [195])

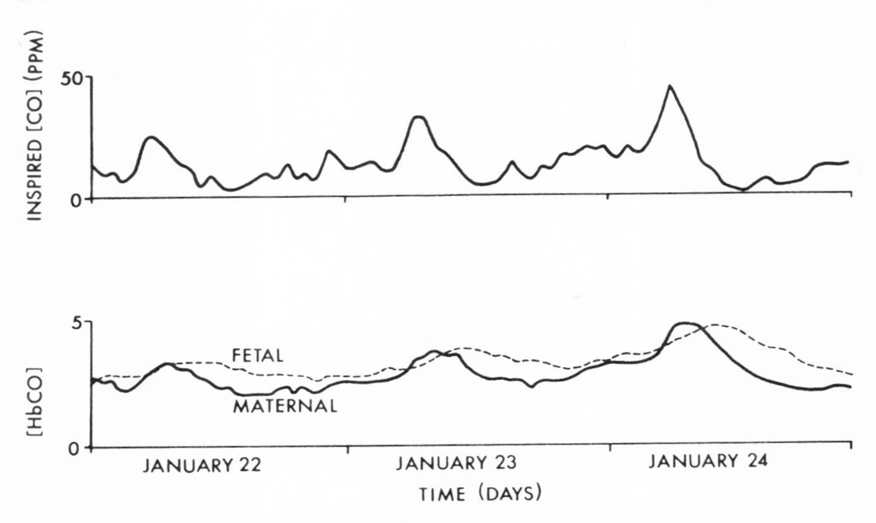

FIGURE 5-13 Measured carbon monoxide concentrations in inspired air (a) and calculated maternal and fetal carboxyhemoglobin concentrations (b) during a 3-day period in southern Los Angeles. Note that fetal carboxyhemoglobin concentrations rise slightly higher than maternal concentrations following each peak in carbon monoxide exposure, and that the fetal carboxyhemoglobin concentrations take longer to decline after the peaks. (Reprinted with permission from Hill *et al.*[195])

if the exposed subject smoked two packs of cigarettes per day, the fetal carboxyhemoglobin concentration may reach 10 to 11%.

MATERNAL SMOKING AND OTHER CARBON MONOXIDE EXPOSURE

Perhaps the most common source of fetal exposure to greater than normal carbon monoxide concentrations is maternal smoking. Several studies have reported the carboxyhemoglobin concentrations in the blood of mothers that smoke and their newborn (Table 5-1). Carboxyhemoglobin concentrations of the fetuses ranged from 2 to 10%, and those of the mothers ranged from 2 to 14%.

The blood samples were obtained at the time of vaginal delivery or cesarean section and probably did not accurately reflect the normal values of carboxyhemoglobin for several reasons: The number of cigarettes smoked during labor may have been less than the number normally consumed; blood samples were collected at varying time intervals following the cessation of smoking; and many samples were probably taken in the morning before the carboxyhemoglobin concentrations had built up to the values reached after prolonged periods of smoking. Therefore, the

TABLE 5-1 Relation of the Concentrations of Fetal to Maternal Carboxyhemoglobin in Mothers Who Smoke during Pregnancy

Fetal Carboxyhemoglobin Concentration, %	Maternal Carboxyhemoglobin Concentration, %	Fetal/Maternal Carboxyhemoglobin Ratio	Reference
7.6 (± 1.14 [SEM])[a]	6.2 (± 0.75)[a]	1.2 (± 0.2)[a]	Haddon *et al.*[179]
3.1 (± 0.84)[b]	3.6 (± 1.06)[b]	0.9 (± 0.14)	
5.0 (± 0.48)	6.7 (± 0.61)	0.7 (± 0.04)	Heron[191]
2.4 (± 0.30)	2.0 (± 0.31)	1.2 (± 0.08)	Young and Pugh[501]
5.3 (± 0.22)	5.7 (± 0.24)	0.9 (± 0.06)	Tanaka[433]
7.3	8.3	0.9	Younoszai and Haworth[502]
3.6 (± 0.7)	6.3 (± 1.7)	0.6 (± 0.15)	Longo[275]
7.5[c]	4.1	1.8	Cole *et al.*[99]

[a] One or more cigarettes 1 hr or less prior to delivery.
[b] One or more cigarettes 1 to 24 hr prior to delivery.
[c] Calculated from $[HbCO_m]$ and the ratio of $[HbCO_f]$ to $[HbCO_m]$.

concentrations measured in both maternal and fetal blood may have been lower than average values for normal smoking periods.

The relation between maternal smoking and low birthweight recently has been reviewed.[59,62,308,309,407] The reports relating perinatal mortality to maternal smoking habits disagree. Several report an increased incidence of spontaneous abortion and of fetal, neonatal, and postneonatal deaths associated with maternal smoking.[59,60,105,146,308,310,368,390,391] Others[350,448,499] reported finding little correlation of these problems with smoking, but this conclusion probably was based on inadequate sample size.

All recent studies using data from large population groups have concluded that perinatal mortality is increased in the infants of mothers who smoke. This increased perinatal mortality is independent of, rather than due to, birthweight reduction.[59,62,106,309,310,330,336] While the mean duration of pregnancy of smoking mothers is slightly shorter than normal,[56,146,281] the proportion of preterm births increased significantly.[59,500] Goldstein[162] has pointed out that the lower perinatal mortality among infants weighing less than 2,500 g born to women that smoke probably reflects the increased mean birthweight of the smokers' babies compared with those of the nonsmokers. Meyer and her colleagues[310] conclude from the Ontario, Canada, data that the independent effect of maternal smoking increases the perinatal mortality risk 20% for light smokers (less than one pack per day) and 35% for heavy smokers (one pack or more per day).

Several causes have been suggested for the low birthweight and increased perinatal mortality: decreased food intake associated with smoking,[389] decreased placental blood flow due to the action of pharmacologic agents in the smoke, and the effects of carbon monoxide on tissue oxygenation. While carbon monoxide probably adversely affects fetal growth and development, other factors such as various chemicals in tobacco smoke and the psychologic makeup of the mother make it difficult to assess the specific effects of carbon monoxide *per se*.

Several reports have analyzed the incidence of complications of pregnancy and labor in smoking mothers. These are increases in the incidence of abruptio placentae with resulting stillbirth,[169,309] placenta previa, and other causes of bleeding during pregnancy.[242,309,337] The incidence of premature rupture of the fetal membranes is also increased, while that of the hypertensive disorders of pregnancy is decreased.[8,59,242,390,448]

Recently, it has been observed that "breathing" movements by the fetus are a normal component of intrauterine development. Both the proportion of time the fetus makes breathing movements and the character of these movements indicate the condition of the fetus. In women with normal pregnancies, cigarette smoking caused an abrupt and significant decrease from a control value of 65% to 50% in the proportion of time that the fetus made breathing movements.[152,288] Carbon monoxide may not play an important role causing these acute changes, however, since marked decreases in breathing were not observed in the fetuses of women who smoked nonnicotine cigarettes.[289] Heron[191] reported a delayed onset of crying immediately after birth in the infants of smoking mothers. Several infants showed definite evidence of asphyxia with irregular respiration and cyanosis.

Long-term effects in surviving children owing to maternal smoking are not well documented. In a multifactorial analysis of data from over 5,000 children in the British National Child Development Study, Davie *et al.*[117] and Goldstein[163] found highly significant differences in reading attainment at 7 yr between the children of mothers who smoked and those who did not. Butler and Goldstein[61] restudied these children at 11 yr of age. Significant differences in the offspring of mothers who smoked 10 or more cigarettes a day included: 3, 4, and 5 months of retardation in general ability, reading, and mathematics, respectively, and a mean height 1 cm less. Their data suggest a decrement in intelligence quotient, but the difference was not statistically significant. Finally, Denson *et al.*[121] reported a syndrome of minimal brain dysfunction in the infants of mothers who smoke. These findings require confirmation as their implications are great.

As noted in other portions of this report, there is a high correlation between smoking and the development of coronary and peripheral artery disease. Several studies indicate that carbon monoxide in tobacco smoke injures arterial blood vessels.[20] Exposure to low doses of carbon monoxide accelerated atherogenesis in cholesterol-fed animals,[21,50,468] producing significant ultrastructural changes in the aortic and coronary epithelium of rabbits[229] and primates,[442,468] indistinguishable from early atherosclerosis. Asmussen and Kjeldsen[17] used the human umbilical artery as a model to evaluate vascular damage caused by tobacco smoking. In comparison with the vessels from babies of nonsmoking mothers, the umbilical arteries from babies of smoking mothers showed pronounced vascular intimal changes. Scanning electromicroscopy disclosed swollen and irregular endothelial cells with a peculiar cobblestone appearance and cytoplasmic protrusions or blebs on their surface. Transmission electromicroscopy showed degenerative changes including endothelial swelling, dilation of the rough endoplasmic reticulum, abnormal-appearing lysosomes, and extensive subendothelial edema. In addition, the basement membrane was markedly thickened, a change probably indicating reparative change. Finally, the vessels showed focal openings of intercellular junctions and loss of collagen fibers. This study underscores the vulnerability of the fetus to the effects of smoking by the mother.

Animal studies showed fetal growth retardation and increased perinatal mortality in pregnant rats[136,503] and rabbits[394] exposed to tobacco smoke. Schoeneck[394] exposed rabbits to tobacco smoke for several generations. The original doe weighed 3.5 kg. One female of the first generation weighed 2.8 kg, another from the second generation weighed only 1.53 kg, and all attempts to breed the doe were either totally unsuccessful or resulted in stillbirths or neonatal deaths.

Of course, factors other than carbon monoxide in tobacco smoke may also result in fetal growth retardation. Younoszai *et al.*[503] exposed rats to several types of smoke, including: the smoke of tobacco leaves, smoke from lettuce leaves plus nicotine, and smoke from lettuce leaves alone. The body weight of rat fetuses exposed to lettuce leaf smoke decreased 9%. Body weight of the fetuses exposed to lettuce leaf smoke plus nicotine decreased about 12%, while the decrease in those animals exposed to tobacco smoke was about 17%. The carboxyhemoglobin concentration was maintained at 2 to 8% in all animals, but the data were not given.

Biologic Effects of Carbon Monoxide on the Developing Embryo, Fetus, and Newborn

EXPERIMENTAL STUDIES OF MAMMALIAN FETAL GROWTH AND SURVIVAL

Few studies have reported the effects of carbon monoxide on fetal growth. Wells[477] exposed pregnant rats to 1.5% (15,000 ppm carbon monoxide) for from 5 to 8 min 10 times on alternate days during their 21-day pregnancy. This resulted in maternal unconsciousness and abortion or absorption of most fetuses. The surviving newborns did not grow normally. Similar exposure to 5,900 ppm affected only a small percentage of animals. This is a brief report lacking quantitative data on the number of experimental animals and number and weight of the fetuses. Williams and Smith[487] exposed rats to 0.34% (3,400 ppm) carbon monoxide for 1 hr daily for 3 months. Peak carboxyhemoglobin concentrations in these animals varied from 60 to 70%. The number of pregnancies known to occur among the seven exposed animals was half the number in the controls. The number of rats born per litter decreased, and only 2 out of 13 newborns survived to weaning age. No pregnancies resulted in the 5 females exposed for 150 days.

Astrup *et al.*[22] reported quantitative data on fetal weights of two groups of pregnant rabbits exposed to carbon monoxide continuously for 30 days. Exposure to 90 ppm resulted in maternal carboxyhemoglobin concentrations of 9 to 10%. Birthweights decreased 11%, from 57.7 to 51.0 g, and neonatal mortality increased to 10%, from a control value of 4.5%. Mortality of the young rabbits during the following 21 days increased to 25% from a control value of 13%. Following exposure to 180 ppm carbon monoxide, with resulting maternal carboxyhemoglobin concentrations of 16 to 18%, birthweights decreased 20% from 53.7 to 44.7 g, and neonatal mortality was 35% compared with 1% for the controls. Mortality during the following 21 days was 27%, the same value as the controls.

CARBON MONOXIDE EFFECTS ON AVIAN EMBRYOGENESIS

Baker and Tumasonis[27] continuously exposed fertilized chicken eggs to various carbon monoxide concentrations from the time they were laid for up to 18 days of incubation. Hatchability correlated inversely with carbon monoxide concentration. At 425 ppm, the apparent "critical level," only about 75% of the eggs hatched. The embryos of these eggs weighed almost the same as those of the controls, and no congenital anomalies were noted. Since 0 and 100 ppm were the only lower carbon monoxide concentrations tested, the conclusion that 425 ppm represents a "critical level" may not be justified. In eggs exposed to 425 ppm, the carboxyhemoglobin concentrations varied from 4 to 16%. The lowest values were reported on the fifteenth to sixteenth day of incubation. At 650 ppm carbon monoxide, the percent of eggs hatching decreased to 46%, and developmental anomalies of the tibia and metatarsal bones were noted. Subsequently, Baker *et al.*[28] exposed embryonated eggs 12 and 18 days old to 425 ppm for 24 hr. The carboxyhemoglobin concentrations averaged 7% in 12- to 14-day chick embryos, 16% in the 16-day embryos, and 36% in the 18-day embryos. It is not clear why the carboxyhemoglobin increased so markedly with embryonic age. The activities of two hepatic mixed-function oxidase enzymes, hydroxylase and *O*-demethylase, increased about 50% in the livers of the older chicks. Chick eggs 13 days old did not show significant enzyme increases. The authors interpreted these results as indicating that the increased hepatic mixed-function oxidase enzymes represent an adaptation of carbon-monoxide-induced tissue hypoxia in the more mature embryo.

EXPOSURE TO EXCESSIVELY HIGH CARBON MONOXIDE CONCENTRATIONS DURING THE NEWBORN PERIOD

Behrman *et al.*[40] reported that, in 16 normal human newborns in a nursery in downtown Chicago, the carboxyhemoglobin concentration increased from about 1% to 6.98 ± 0.55% and the blood oxygen capacity decreased 13.8 ± 0.57% (control values were not given). The authors correlated the decreases of blood oxygen capacity with carbon monoxide concentrations at a Chicago air pollution station about 1.5 miles distant. On days when the ambient carbon monoxide concentration was less than 20 ppm, blood oxygen capacity was decreased 8.4 ± 1.1% in infants aged 24 hr old or younger and 11.0 ± 0.73% in infants over 24 hr old. On days when the atmospheric carbon monoxide was greater than 20 ppm, blood oxygen capacity decreased 11.4 ± 1.1% in the newborns and 13.0 ± 0.65% in babies over 24 hr old. The carboxyhemoglobin concentrations varied as a function of both the concentrations of inspired carbon monoxide and the duration of exposure. While these observations are of interest, certain problems inherent in the study were not clarified. A close correlation between the carbon monoxide concentrations in the nursery and at the monitoring station 1.5 miles distant is questionable. The authors noted that the decrease in oxygen capacity was greater than could be accounted for by carbon monoxide alone, although the error in the method of carboxyhemoglobin determination with the spectrophotometer used probably invalidates such a conclusion. No untoward clinical effects were observed either from these carboxyhemoglobin concentrations or from the decrease in blood oxygen capacity.

Smith *et al.*[453] studied the effects of carbon monoxide on newborn survival in animals. They exposed rats to mixtures of illuminating gas in air, with inspired carbon monoxide concentrations of 0.43%. In 22 newborn rats 12 to 48 hr old exposed to carbon monoxide, the average survival time was about 196 min, in contrast to an average survival of about 36 min in mature animals. McGrath and Jaeger[301] also noted that 50% of newly hatched chicks could withstand exposure to 1% (10,000 ppm) carbon monoxide concentration for about 32 min. This initial resistance to carbon monoxide decreased rapidly. By day 1, mean survival time decreased to about 10 min, by day 4 it was 6 min, and by day 8 it was 4 min, where it remained for all ages tested up to 21 days. Subsequently, Jaeger and McGrath[209] showed that decreasing the body temperature increased the time to last gasp from a mean value of 9.8 ± 0.5 min at 40 C to 20.7 ± 0.1 min at 30 C. They noted that hypothermia caused markedly reduced heart and respiratory rates and suggested that its major benefit was a reduction in energy-requiring functions.

POSSIBLE MECHANISMS BY WHICH CARBON MONOXIDE AFFECTS THE FETUS AND NEWBORN

Several mechanisms probably account for the effects of carbon monoxide on developing tissue. Undoubtedly the most important of these is the interference with tissue oxygenation.[30,144] As first observed by Claude Bernard in 1857,[48] carbon monoxide decreases the capacity of blood to transport oxygen by competing with it for hemoglobin. When carbon monoxide binds to hemoglobin, the oxygen affinity of the remaining hemoglobin is increased. This shift to the left of the oxyhemoglobin saturation curve means that the oxygen tension of blood must decrease to lower than normal values before a given amount of oxygen will be released from hemoglobin. This effect may be particularly significant for the fetus because the oxygen tension in the arterial blood is normally relatively low, about 20–30 mm Hg, as compared to adult values of about 100 mm Hg. Carbon monoxide also interferes with oxygen transport by displacing oxygen from the hemoglobin in arterial blood, thus decreasing the blood oxygen capacity. For the pregnant woman, these effects on blood oxygenation pose a special threat. Not only is her oxygen consumption increased 15–25% during pregnancy,[348] but her blood oxygen capacity is decreased 20 to 30% or more due to the decreased concentration of hemoglobin.[190] The woman with a significant anemia faces even more severe compromise of her oxygen delivery.

The theoretical basis for understanding the consequences of carbon monoxide interaction with oxygen in humans is shown in Figure 5-14, in which blood oxygen content is plotted as a function of the oxygen partial pressure. The oxygen affinity of fetal blood is greater than that of maternal blood, hence its oxyhemoglobin saturation curve is shifted to the left. In addition, human fetal blood contains more hemoglobin than maternal (16.3 vs. 12 g/100 ml); therefore it has a greater oxygen

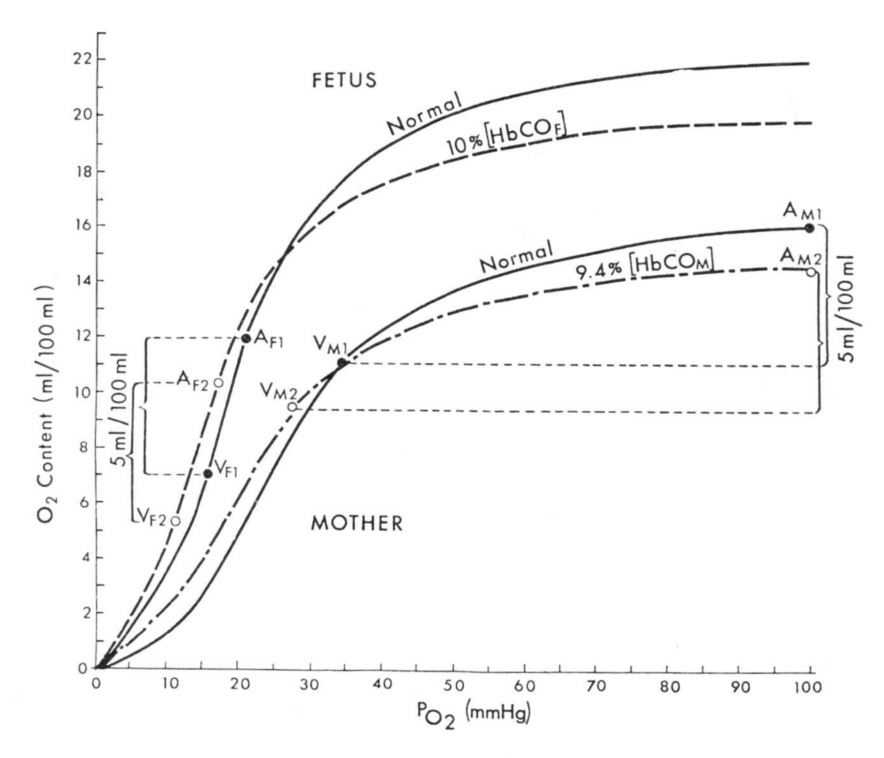

FIGURE 5-14 Oxyhemoglobin saturation curves (plotted as blood oxygen content vs. partial pressure) of human fetal blood with 0 and 10% carboxyhemoglobin and of maternal blood with 0 and 9.4% carboxyhemoglobin concentrations. This figure depicts the mechanism accounting for the reduction of umbilical artery and vein oxygen partial pressures and contents resulting from elevated carboxyhemoglobin concentrations. (See text for details.) (Reprinted with permission from Longo.[274] Copyright 1976 by the American Association for the Advancement of Science.)

capacity.[190] Under normal circumstances, maternal arterial oxygen tension is about 100 mm Hg and its oxygen content equals 16.1 ml/100 ml of blood (point A_{M1} in Figure 5-14). The placental exchange of oxygen extracts about 5 ml O_2/100 ml blood, producing a uterine mixed venous oxygen tension of about 34 mm Hg (point V_{M1}). When the maternal blood contains carboxyhemoglobin, the oxygen capacity is decreased and the oxyhemoglobin curve shifts to the left (as indicated by the curve labeled 9.4% [$HbCO_m$]). While arterial oxygen tension remains essentially the same as under normal conditions, the oxygen content is reduced to 14.5 ml/100 ml (point A_{M2}). With the same placental oxygen transfer of 5 ml/100 ml blood, the venous oxygen tension would be about 27 mm Hg (point V_{M2}), a decrease from normal of about 7 mm Hg.

In the fetus, oxygen partial pressure in the descending aorta is normally about 20 mm Hg, and oxygen content is about 12 ml/100 ml (point A_{F1}). With an oxygen consumption of 5 ml/100 ml, inferior vena caval oxygen tension would be 16 mm Hg (point V_{F1}). With elevated fetal carboxyhemoglobin concentrations, both arterial (point A_{F2}) and venous (point V_{F2}) oxygen tensions are reduced. This contrasts to the adult, in whom the arterial oxygen tension remains normal. This is because of the lowered oxygen tension of maternal placental capillaries with which fetal blood equilibrates. With normal fetal oxygen consumption, the venous blood oxygen content is 5 ml/100 ml less than arterial content, producing a venous oxygen tension of about 11 mm Hg, a decrease from normal of 5 mm Hg.

The oxygen tension of venous blood is roughly indicative of the adequacy of tissue oxygenation, and the mean capillary partial pressure driving oxygen into the tissues is probably related to the oxygen partial pressure at 50% oxyhemoglobin saturation, the p50. Figure 5-15 shows the changes

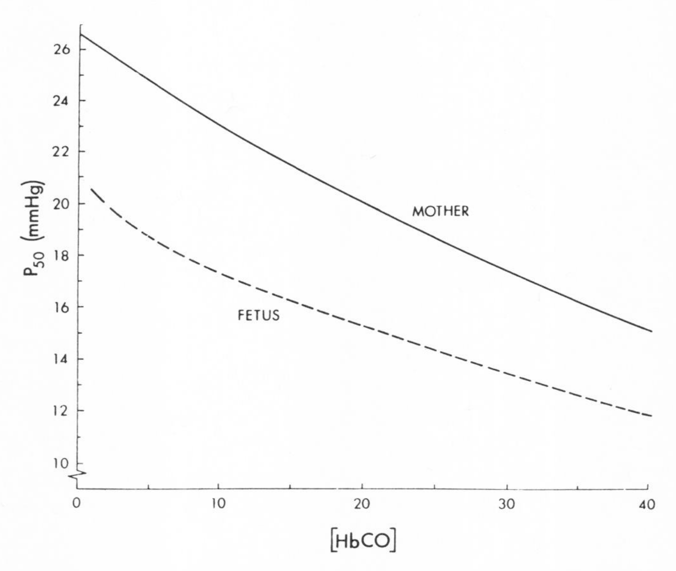

FIGURE 5-15 The partial pressure at which the oxyhemoglobin saturation is 50% (p50) for human maternal and fetal blood as a function of blood carboxyhemoglobin concentration. (Reprinted with permission from Longo.[276])

in p50 for maternal and fetal blood as a function of the blood carboxyhemoglobin concentrations.

While the effects of carbon monoxide on venous oxygen tensions have been considered from a theoretical standpoint, there has been essentially no experimental validation of these effects in either adults or the fetus. Longo and Hill[278] recently examined the changes in oxygen tension in response to various carboxyhemoglobin concentrations in sheep in which catheters were chronically implanted in maternal and fetal vessels. About 1 week following their recovery from anesthesia and surgery, the ewes were exposed to various concentrations of carbon monoxide.

Figure 5-16 shows the oxygen tensions in the descending aorta and the inferior vena caval below the ductus venosus as a function of carboxyhemoglobin concentration in the fetus. Maternal and fetal carboxyhemoglobin levels were in quasi-steady-state equilibrium. In contrast to the adult, in which arterial oxygen tension is relatively unaffected by changes in carboxyhemoglobin concentrations, fetal arterial oxygen tension is particularly sensitive to increases in maternal or fetal carboxyhemoglobin concentrations. This is because fetal arterial oxygen tension varies with the oxygen tension in fetal placental end-capillary blood, which in turn varies with the oxygen tensions in maternal placental exchange vessels. The oxygen partial pressure in the fetal descending aorta decreased from a control value of about 20 mm Hg to 15.5 mm Hg at 10% fetal carboxyhemoglobin concentration (Figure 5-16). The regression equation for this relation was: $pO_2 = 20.1 - 0.4\,[HbCO_f]$ ($R = -0.96$).

Figure 5-16 also shows the relation of oxygen tension of the inferior vena cava, below the ductus venosus, to carboxyhemoglobin concentration of the fetus. At 10% carboxyhemoglobin concentration, inferior vena caval oxygen tension decreased from a control value of about 16 to 12.5 mm Hg. The regression equation for this relation was: $pO_2 = 15.8 - 0.3\,[HbCO_f]$ ($R = -0.96$). Strictly speaking, the oxygen tension of fetal venous blood is affected both by the decreased maternal placental venous oxygen tension resulting from increased maternal carboxyhemoglobin concentration and the increased fetal carboxyhemoglobin concentration. Since these oxygen tensions were obtained when maternal and fetal carboxyhemoglobin concentrations were in a quasi-steady-state condition, a relation between fetal inferior vena caval oxygen tension and maternal carboxyhemoglobin concentration would be expected. If one plots this relation,[274,278] the decrements in fetal oxygen tension appear greater when plotted as a function of maternal carboxyhemoglobin than when plotted as a function of fetal carboxyhemoglobin concentration. This follows because the carboxyhemoglobin concentration of the fetus exceeds that of the mother during equilibrium conditions.

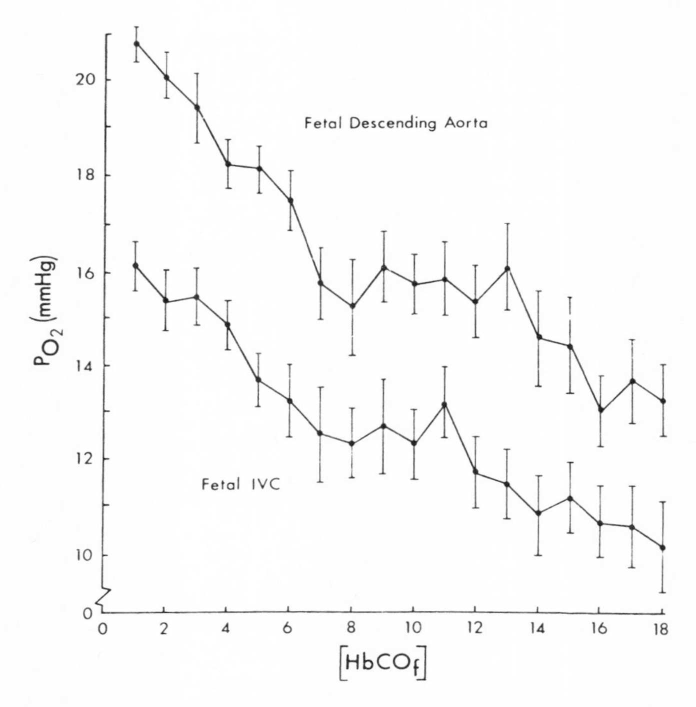

FIGURE 5-16 Fetal values of oxygen partial pressure as a function of carboxyhemoglobin concentrations during quasi-steady-state conditions. Fetal inferior vena caval oxygen tension is a function of both maternal and fetal carboxyhemoglobin concentrations. The oxygen partial pressure of fetal arterial blood is chiefly a function of maternal carboxyhemoglobin concentrations. During steady-state conditions, however, it is also related to the fetal carboxyhemoglobin concentration. Each point represents the mean ± SEM (vertical bars) of 6–20 determinations at each concentration of blood carboxyhemoglobin. (Reprinted with permission from Longo.[274] Copyright 1976 by the American Association for the Advancement of Science.)

These oxygen partial pressures in sheep are not identical with those values anticipated in humans because of differences in oxygen affinities and capacities of maternal and fetal blood between the species. Differences, however, are estimated to be no more than 2–3 mm Hg.

About 57% of the sheep fetuses in this study died when fetal carboxyhemoglobin values were greater than 15% for 30 min or longer (5 of 11 died at 100 ppm, and 3 of 3 died at 300 ppm).[274] These deaths presumably resulted from hypoxia of vital tissues. There are probably two major reasons for this. First, in the adult, elevation of carboxyhemoglobin concentrations to 15–20% results in a 6–10 mm Hg decrease in venous oxygen tension. While this decrease is substantial, the resultant oxygen partial pressures probably remain well above critical values for maintaining oxygen delivery to the tissues.[415] In contrast, in the fetus with normal arterial and venous oxygen tension probably close to the critical levels, a 6–10 mm Hg decrease in oxygen tension can result in tissue hypoxia or anoxia. Furthermore, adult subjects and animals subjected to carbon monoxide hypoxia show increases in cardiac output,[25] coronary blood flow (R. Gilbert, S. Carlson, and B. Bromberger-Barnea, unpublished paper), and presumably tissue blood flow. Apparently, such compensatory adjustments are not available to the fetus to any great extent. The decreases in blood oxygen tensions measured experimentally were similar to those predicted, assuming no increase in tissue blood flow. In addition, the fetus probably cannot increase its cardiac output significantly, as fetal cardiac output normally is about two to three times that of the adult on a per weight basis.[356] Thus, the fetus probably normally operates near the peak of its cardiac function curve.

Ginsberg and Myers[155,156] studied the effects of carbon monoxide exposure on near-term pregnant monkeys and their fetuses. They acutely exposed anesthetized animals to 0.1 to 0.3% carbon monoxide, resulting in maternal carboxyhemoglobin concentrations of about 60%. During the 1- to 3-hr studies, fetal blood oxygen content decreased to less than 2 ml/100 ml blood. Fetal heart rate decreased in proportion to the blood oxygen values. These fetuses also developed severe acidosis (pH less than 7.05), hypercarbia (pCO_2 = 70 mm Hg or greater), hypotension, and electrocardiographic changes such as T-wave flattening and inversion.[155]

In cases of acute carbon monoxide poisoning, the mother will develop a high carboxyhemoglobin concentration with a shift to the left of her oxyhemoglobin saturation curve, while the carboxyhemoglobin concentration in the fetus remains normal or low. Nonetheless, the fetus experiences a decrease in pO_2 values because of the shift in the maternal dissociation curve.[155,277]

In a theoretical analysis, Permutt and Farhi[347] calculated the effect of elevated carboxyhemoglobin in lowering oxygen tension in the venous blood of an otherwise normal adult. Hill *et al.*[195] used this approach to calculate the effect of various carboxyhemoglobin concentrations on fetal tissue oxygenation. The equivalent reduction in umbilical arterial oxygen tension required to maintain normal placental oxygen exchange or the equivalent increase in blood flow necessary for oxygen exchange was calculated. The results are plotted in Figure 5-17. A 10% carboxyhemoglobin concentration would be equivalent to a 27% reduction in

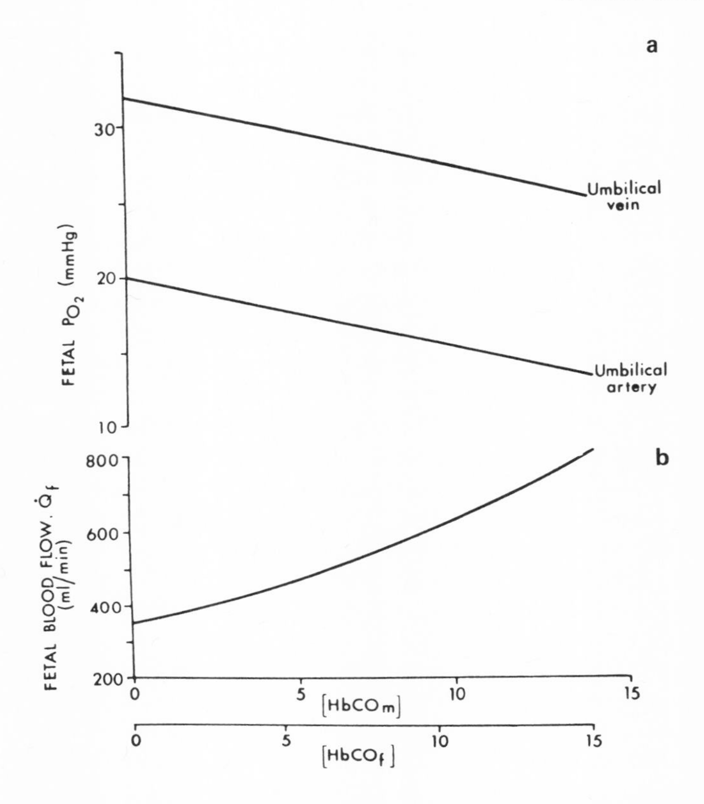

FIGURE 5-17 (a) Umbilical arterial oxygen partial pressure drop necessary to maintain normal oxygen exchange at various carboxyhemoglobin concentrations allowing no change in maternal or fetal blood flows or maternal arterial oxygen partial pressures. (b) Increase in fetal blood flow necessary to maintain oxygen exchange (with no change in umbilical arterial oxygen tension). These curves indicate the degree of compensation necessary to offset the effects of elevations in carboxyhemoglobin concentrations. (Reprinted with permission from Hill *et al.*[195])

umbilical arterial pO_2 (from 18 to 13 mm Hg). It would also be equivalent to a drastic reduction in blood flow. Fetal blood flow would have to increase 62% (from 350 to 570 ml/min) to maintain normal oxygen exchange. Higher levels of fetal carboxyhemoglobin require even more dramatic compensations.

A further carbon monoxide effect is its relation to oxygen consumption by placental tissue. Tanaka[433] used a Warburg apparatus to measure oxygen consumption in placental slices from nonsmoking and smoking mothers. With tissues from normal nonsmoking mothers, the oxygen consumption was about 1.9 μl/mg placenta per hr. With the tissue from smoking mothers, the rate of oxygen consumption decreased in proportion to the concentration of carboxyhemoglobin in the maternal blood. For example, it decreased about 30% to 1.3 μl/mg per hr at a maternal carboxyhemoglobin concentration of 8%.

Hypoxia owing to carbon monoxide may further interfere with tissue oxygenation by increasing the concentration of carboxymyoglobin, [MbCO], in relation to carboxyhemoglobin. As shown by Coburn and his co-workers, when arterial oxygen tension decreases to 40 mm Hg or lower, the ratio of carboxymyoglobin to carboxyhemoglobin increases from a normal value of 1.0 to about 2.5 in the resting dog skeletal muscle[93] and myocardium[94] of dogs. If a similar relation exists in the fetus, where arterial pO_2 is normally 30 mm Hg or less, then carboxymyoglobin concentration would be 12.5% when carboxyhemoglobin is 5%. The effect of this degree of carboxymyoglobin saturation on oxygenation of the fetal myocardial cells may be significant.

SIMILARITIES OF HIGH-ALTITUDE HYPOXIA AND CARBON MONOXIDE HYPOXIA

The interference with tissue oxygenation caused by carbon monoxide is somewhat similar to hypoxia at high altitudes. Both can result in lower oxygen tensions in end-capillary blood and therefore tissue hypoxia. The effect of smoking on newborn birthweight is strikingly similar to the effects of high altitude. A study by the Department of Health, Education, and Welfare reported in 1954 that in the Rocky Mountain states mean birthweights were lower, there were a larger proportion of babies weighing less than 2,500 g and fewer weighing greater than 4,000 g, and neonatal mortality was higher than in the country as a whole.[401] A comparison of infants born in Lake County, Colorado, elevation about 3,000–3,600 m (9,800–11,800 ft), with those born in Denver, elevation about 1,585 m (5,200 ft), showed that mean birthweight of the Lake County babies was 290 g less than that of those born in Denver.[261] In Lake County, 48.3% of newborns weighed less than 2,500 g compared with 11.7% in Denver.

A more precise analysis of the relation between altitude and birthweight can be derived from the work of Grahn and Kratchman.[170] These workers showed that the proportion of low birthweights increased with increasing elevation. The number of infants weighing less than 2,500 g increased to 10.1% at 1,500–1,580 m (4,920–5,180 ft) from a control value of 6.6% at sea level. At 2,760 m (9,055 ft), the number of such newborns further increased to 16.6%. The decrease in the duration of gestation, a mean difference of 0.4 week, was not enough to explain the 190 g birthweight difference between Colorado and Illinois–Indiana infants.[170]

A given carboxyhemoglobin concentration may cause a more profound effect on tissue oxygenation at high altitudes than at sea level. Thus, individuals at high altitudes would be expected to be more sensitive to the effects of carbon monoxide. This would be particularly true of the pregnant mother, whose oxygen requirement is greater than when not pregnant,[348] as well as for the newborn infant.

CARDIOVASCULAR RESPONSES TO CARBON MONOXIDE EXPOSURE

The heart requires a continuously available supply of oxygen in order to maintain electrical and contractile integrity. The glycolytic response to hypoxia in the myocardium is much less than that observed in skeletal muscle and is unable to prevent a rapid depletion of high-energy phosphates. Ectopic electrical activity, decline in contractile force, and ventricular fibrillation soon follow the induction of myocardial hypoxia. As discussed above, carbon monoxide decreases the oxygen-carrying capacity of hemoglobin and shifts the oxyhemoglobin dissociation curve to the left, reducing the oxygen tension in both the capillaries and cells (see Figure 5-18).

The concept that any alteration in intracellular oxygen tension is rapidly corrected by an appropriate increase or decrease in coronary blood flow is central to an understanding of the effects of carboxyhemoglobinemia on myocardial oxygenation. Peripheral tissues respond to increased oxygen needs by increasing oxygen extraction and reducing venous oxygen content and tension. Increased myocardial oxygen extraction is almost never seen in the normal heart and, when present, is considered evidence of myocardial hypoxia.

The capacity for increased coronary blood flow in the normal heart is demonstrated by increases of up to 300% in the coronary flow during exercise, in severe anemia,[173] or in individuals exposed to arterial hypoxemia.[327] The myocardium probably also adapts to hypoxia by increasing "functioning" capillary density.[196] The capacity of the normal heart to maintain its oxygenation at relatively high carboxyhemoglobin concentrations is analogous to the situation in which a normal man climbs to high altitudes or a patient with chronic anemia sustains a 70% reduction in circulating hemoglobin.

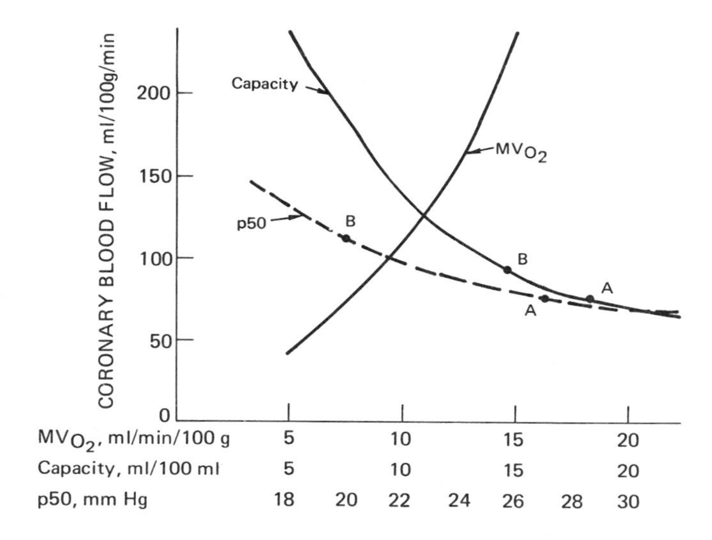

FIGURE 5-18 Analogue model illustrating theoretical relationships between coronary blood flow and myocardial oxygen consumption (MV_{O_2}), hemoglobin oxygen-carrying capacity (Capacity), and the position of the oxyhemoglobin dissociation curve (p50). Point A on the p50 and capacity curves represents coronary blood flow at normal p50 and hemoglobin capacity; Point B illustrates theoretical effect of increasing carboxyhemoglobin from 0 to 20% saturation on coronary blood flow. The assumptions used in computing this theoretical model are given by Duvelleroy.[132] (Adapted from Duvelleroy *et al.*[132])

The importance of coronary blood flow in the maintenance of myocardial oxygen tension is emphasized in a recent theoretical model presented by Duvelleroy *et al.*[132] Increasing myocardial oxygen consumption (by increasing heart rate, arterial blood pressure, or contractile force), decreasing oxygen capacity, or shifting the oxyhemoglobin dissociation curve to the left all lead to a decrease in myocardial oxygen tension unless coronary blood flow is increased commensurately. Figure 5-18 shows that each alteration can independently produce a change in coronary blood flow which restores myocardial oxygen tension to normal. Increasing carboxyhemoglobin concentrations increase coronary blood flow both by decreasing the oxygen capacity of hemoglobin and by shifting the oxyhemoglobin curve to the left, shown in the figure as a decrease in the oxygen tension at 50% concentration, p50. Points A and B show the effects of 0 and 20% carboxyhemoglobinemia, respectively. Note that the relationships are curvilinear so that progressively greater alterations call forth disproportionately greater increases in coronary blood flow. The implications of this type of response curve for elevated baseline conditions are discussed elsewhere in this chapter. The cardiovascular response of the intact organism to carbon monoxide inhalation depends on the ability of the entire coronary vascular bed to dilate and increase coronary blood flow. The random occurrence of coronary atherosclerosis in the general population explains the variation in myocardial responses observed both in humans and among different species. It also provides a physiologic explanation for the clinical observations described in the next section.

The manifestations of carbon monoxide toxicity were shown to be closely related to the carboxyhemoglobin concentration by John Haldane[183] in 1895. He did not observe serious symptoms at rest until his own hemoglobin was at least one-third saturated with carbon monoxide. Exertion, however, produced mild dyspnea and palpitations when as little as 14% carboxyhemoglobin was present. Haggard[181] in 1921 and Chiodi and co-workers[79] in 1941 observed increase in pulse rate and cardiac output with carboxyhemoglobin concentrations between 16 and 20%.

Ayres *et al.*[25] studied the systemic hemodynamic and respiratory response to acute increases in carboxyhemoglobin in man by means of measurements performed during diagnostic cardiac catheterization. Carboxyhemoglobin concentrations between 6 and 12% were achieved by the breathing of either 5% carbon monoxide for 30–45 s or 0.1% carbon monoxide for 8–15 min. In men with no evidence of heart disease, cardiac output increased from 5.01 to 5.56 l/min, the minute ventilation increased from 6.86 to 8.64 l/min, and arterial carbon dioxide pressure, pCO_2, decreased from 40 to 38 mm Hg. Systemic oxygen extraction ratios increased from 0.27:1 to 0.32:1, indicating more complete extraction of oxygen from perfusing arterial blood. Mixed venous oxygen tension decreased from 39 to 31 mm Hg as a result of the leftward shift of the oxyhemoglobin dissociation curve and arterial pO_2 unexpectedly decreased from an average of 81 to 76 mm Hg. The decrease is probably due to enhancement of the venoarterial shunt effect and is more prominent in patients who initially have a low arterial pO_2.

Observations such as this suggest that carbon monoxide inhalation would have a significant effect on arterial pO_2 in those patients with preexisting lung disease, in those individuals who are heavy smokers, and in patients in coma owing to severe carbon monoxide poisoning. These studies were repeated with the lower concentration of carbon monoxide (0.1% for 8–15 min). Cardiac output did not change significantly, although pCO_2 decreased, indicating hyperventilation. Changes in arterial and mixed venous pO_2 were similar to those observed with the higher concentrations.

Myocardial studies were performed before and after the administration of either 5 or 0.1% carbon monoxide for 30–45 s or 8–15 min, respectively. Patients were divided into two groups, those with coronary arterial disease and those with other cardiopulmonary disorders. In the patients with other cardiopulmonary diseases, the lowest concentration of carbon monoxide decreased the myocardial arteriovenous oxygen difference by an average of 6.6% and in the patients with coronary arterial disease by 7.9%. The higher concentration decreased the difference by 25 and 30.5%, respectively. Coronary blood flow increased in all but two of the studies, regardless of the dose delivered. These changes were statistically significant.

Neither the presence of coronary arterial disease nor the concentration of carbon monoxide appeared to alter this response to increasing carboxyhemoglobin. Failure of the increase in coronary blood flow to compensate for the decrease in oxygen delivery produced by carbon monoxide was suggested by a decrease in coronary sinus oxygen tension in all but two of the twenty-six patients studied. Nine of fourteen patients with coronary arterial disease had decreased lactate extraction with increasing carboxyhemoglobin, a finding suggestive of anaerobic metabolism. In four of these patients, lactate extraction ceased and the myocardium produced lactate, indicating severe anaerobic metabolism. The increase in their carboxyhemoglobin concentration averaged 5.05%. These studies were performed in patients with arteriographically demonstrated coronary arterial disease. The abnormal changes occurred while the patients were at rest.

Adams *et al.*[2] also found that low concentrations of carboxyhemoglobin increased coronary blood flow. These workers exposed conscious dogs to carbon monoxide at increasing concentrations and observed a 13% increase in coronary blood flow with a 4% increase in carboxyhemoglobin. At 20% carboxyhemoglobin, coronary blood flow had increased by 54%. Coronary blood flow measurements at four carboxyhemoglobin concentrations suggested a linear relation between carboxyhemoglobin concentration and blood flow. A threshold value was not observed, although measurements were not made below 5% carboxyhemoglobin.

Studies in experimental animals with presumably normal coronary vascular beds are useful models for the effects of carbon monoxide on individuals without coronary disease. Since myocardial oxygen tension is maintained by increasing coronary blood flow, in these studies dose-response relationships cannot be directly used for establishing threshold levels for man.

An interesting study of the effect of relatively low concentrations of carbon monoxide in the cynomolgus monkey has been reported by DeBias *et al.*[440] They observed that chronically increased carboxyhemoglobin (to an average of 12.4%) produced polycythemia with an increase in hematocrit from 35 to 50% of the volume of packed red blood cells. All animals developed increased amplitude of the P-wave in the electrocardiogram and some developed T-wave inversions suggestive of myocardial ischemia. Electrocardiographic abnormalities were more severe in the monkeys with experimental myocardial infarction and increased carboxyhemoglobin than in those with experimental infarction who were breathing ambient air.

In a later study, DeBias *et al.*[440] showed that increasing the carboxyhemoglobin concentration to an average of 8.5% in five cynomolgus monkeys reduced the ventricular fibrillation threshold. Fibrillation could be produced by an average of 45 V AC delivered for 150 ms during the vulnerable period compared to 79 V while breathing ambient air ($P < 0.01$). This observation in normal monkeys is particularly relevant to the problem of sudden death in atheromatous man.

The study of ultrastructural changes following exposure is another approach to the evaluation of the myocardial toxicity of carbon monoxide. Thomsen and Kjeldsen[443] observed myofibrillar degeneration and myelin body formation in the mitochondria of rabbits exposed to carbon monoxide at 100 ppm for 4 hr, an exposure sufficient to raise carboxyhemoglobin concentrations to 8 or 9%.

While the currently available physiologic data are insufficient by themselves to formulate dose–response data, they emphasize the multifarious nature of the human response. Sudden exertion in an individual with carboxyhemoglobinemia requires a substantial increase in coronary blood flow in order to overcome the effects of increased myocardial oxygen requirements, decreased oxygen-carrying capacity of hemoglobin, and rightward shift of the oxyhemoglobin dissociation curve (Figure 5-18). Should an appropriate increase in blood flow be limited in any region of the myocardium by a rigid vascular bed, myocardial hypoxia may occur. Such a multivariate formulation emphasizes the difficulty of identifying a single threshold concentration capable of protecting the entire population.

RELATION BETWEEN CARBON MONOXIDE AND CORONARY ARTERIAL DISEASE

Arteriosclerotic heart disease (ASHD) is the leading cause of death in the United States,[233] with approximately 35% of all deaths directly attributable to this disease. It is also a major cause of morbidity. Some of its clinical and epidemiologic characteristics are particularly pertinent for establishing a causal relationship between carbon monoxide exposure and ASHD. The pathologic basis of the clinical disease is a severe, diffuse narrowing of the coronary arteries.[375] Neither the frequency nor the prevalence of coronary arterial stenosis in the population is known. There are, however, many people with asymptomatic severe coronary arterial stenosis due to atherosclerosis. They are at high risk with respect to developing clinical ASHD.

The underlying atherosclerosis is considered to be a chronic disease, but the clinical manifestations are usually acute. Approximately 25–30% of the individuals die suddenly, from several minutes to 24 hr after their first heart attack.[243] Certain risk factors for clinical disease have been identified.[359] They appear to act primarily in the development of the severe underlying atherosclerotic disease and the consequent coronary stenosis. Very little is known about the specific factors that precipitate the clinical disease, such as myocardial infarction, angina pectoris, and sudden death. Agents that decrease the available oxygen supply to the myocardium, including carbon monoxide, are primary suspects as precipitating factors in heart attacks.

Most persons who experience either angina or myocardial infarction have severe coronary arterial stenosis.[349] They are at high risk compared to the rest of the population for recurrent heart attacks and sudden death.[51] About 20% of the men who survive the first months after a myocardial infarction will die within 5 yr.[470] Most of these deaths will be due to recurrent heart attacks, and about 60% of the deaths will be sudden. The specific environmental and host factors that determine survival following a heart attack have not been clearly determined.

Environmental exposure to carbon monoxide and ASHD may be related in two ways. One of these is that exposure to carbon monoxide can enhance the development of underlying atherosclerosis and subsequent coronary arterial stenosis when associated with other risk factors such as increased cholesterol levels and hypertension. The other is that, in the presence of severe underlying coronary arterial stenosis due to atherosclerosis, carbon monoxide may be a major precipitant of myocardial infarction, angina pectoris, or sudden death. Such a relation is very plausible in the light of what is known about the epidemiology of heart disease and the possible pathophysiologic mechanisms.

The relation between carbon monoxide and the precipitation of heart attacks can be extended further to include new cases among individuals with preexisting but silent underlying atherosclerosis; an increase in morbidity, such as greater frequency and severity of chest pain, in patients with angina pectoris; or a reduction in survival among individuals with

TABLE 5-2 Relation between Atherosclerosis and Carbon Monoxide

Animal	Exposure		Results	Reference
	Experimental	Control		
Rabbit	CO at 0.009%	Ambient air	Increased focal degenerative and reparative changes in intimal and subintimal coats in CO-exposed animals	Wanstrup *et al.*[465]
Rabbit	Cholesterol and 170 ppm CO for 7 weeks	Cholesterol and ambient air	Cholesterol content of aorta 2.5 times higher than in control	Astrup *et al.*[21]
Primates and squirrel monkey	Atherogenic diet plus 200–300 ppm CO for 4 hr, 5 days per week, for 7 months	Atherogenic diet only and compressed air	Enhanced atherosclerosis in monkey exposed to carbon monoxide	Webster *et al.*[468]
Primates: *Macaca irus*	250 ppm CO continuously; 250 ppm every 12 hr	Ambient air	Experimental group subendothelial edema and infiltration of cells, lipid droplets in coronary arteries	Thomsen[442]
Rabbit	180 ppm CO for 2 weeks	Air	Exposed, local areas of partial or total necrosis of myofibrils and degenerative changes of mitochondria	Kjeldsen *et al.*[231]

clinical atherosclerotic heart disease, such as decreased life expectancy following a myocardial infarction.

Relation between Exposure to Carbon Monoxide, Severity of Coronary Atherosclerosis, and Possible Precursor Vascular Lesions

Studies of the association between underlying coronary atherosclerosis and exposure to carbon monoxide have been limited to laboratory investigations with animal models (Table 5-2).[21,231,442,465,468] In these studies the experimental animals are exposed to various concentrations of carbon monoxide either continuously or intermittently, while the controls breathe ambient air. For some of the experiments both the experimental and the control animals are fed a diet high in cholesterol and/or fat, and either the cholesterol content of the artery or the extent of vascular disease is measured. Several studies in rabbits and primates have reported that animals exposed to relatively high doses of carbon monoxide (170–180 ppm) for extended periods of time have either a higher cholesterol content in their arteries or enhanced vascular disease. Animals in another type of study were exposed to large doses of carbon monoxide but were not fed a high-cholesterol or high-fat diet. These studies in both primates and rabbits have reported finding subendothelial edema and gaps between the endothelial cells with increased infiltration of the cells with lipid droplets. (These lesions might be the early precursors of atherosclerotic disease.)

There are no studies in humans describing the relation between exposure to carbon monoxide and the rate of development of atherosclerotic disease. Evidence that such a relation exists is based on the observation that cigarette smokers, who have higher carboxyhemoglobin concentrations than nonsmokers, also have more advanced atherosclerosis than nonsmokers.[373] There is also evidence, however, that exposure to carbon monoxide may not be causally related to the underlying atherosclerosis. Heavy cigarette smokers in countries such as Japan, where the diets are low in fat and cholesterol, do not seem to have a high risk of heart attacks and probably do not have severe coronary atherosclerosis.[220]

The enhancement of atherosclerosis due to carbon monoxide exposure might take place only in the presence of a high-fat, high-cholesterol diet. Since much of the United States population eats such a diet, exposure to carbon monoxide could be an important determinant of the extent of coronary atherosclerosis. However, because the data for such an association are preliminary in nature, any current conclusions would be speculative.

TABLE 5-3 Relation between Ambient Carbon Monoxide Levels and Incidence and Survival of Coronary Heart Disease

Study	Method	CO Range	Results
Los Angeles	Comparison of admissions and case fatality from heart disease in high- and low-CO areas	5–14 ppm weekly basin average	No difference in admission rates
	Above or below 8 ppm CO; 2,484 admissions in high areas; 596 in low areas		Increased fatality rates in high-CO areas only during times of high pollution (8.6 + mean weekly ppm CO)
Baltimore	Incidence of sudden death, myocardial infarction, total ASHD deaths compared with ambient CO concentrations at one station	0–4 ppm compared to 9 + ppm CO 24-hr average	No relation between the incidence of sudden death, myocardial infarction, or total ASHD deaths and the ambient CO concentrations

Relation between Exposure to Carbon Monoxide and the Incidence of Clinical Arteriosclerotic Heart Disease

The relation between ambient carbon monoxide concentrations and the incidence of clinical disease in man has been studied in two ways. One of these was by comparing the spatial or temporal distribution of ambient carbon monoxide concentrations with the occurrence of new cases (Table 5-3). The other was by measuring the postmortem carboxyhemoglobin concentrations among ASHD sudden-death subjects and those of subjects who died suddenly from other causes and comparing the values with those of normal living controls.

In two community studies (Los Angeles[161] and Baltimore[244]), a relation between the incidence of heart attack and ambient carbon monoxide concentrations has not been demonstrated. The Los Angeles study was based on hospital admissions for myocardial infarctions, and the Baltimore study was based on sudden and more prolonged deaths due to ASHD and patients admitted to the hospital because of their first transmural myocardial infarction.

The major problem with this type of study is the difficulty in determining the dose of carbon monoxide that a heart attack subject may have received owing to exposure to the ambient carbon monoxide concentrations in the community. The onset of the acute event, myocardial infarction, or sudden death is not well-defined except perhaps for instantaneous deaths.[373] For many myocardial infarction cases and sudden deaths, there is a prodromal period that may last for several days prior to hospitalization or death. Therefore, the ambient carbon monoxide concentrations on the day of admission or death may have little relation either to the onset of the myocardial infarction or to sudden death. Can the values reported by a field-monitoring station be used to estimate the dose or change in dose for an individual living in the community but not in the immediate vicinity of the sampling station? Perhaps it would be advisable to consider the reported values as a crude estimate of changes in exposure from day to day rather than as an individual dose.

Comparing carboxyhemoglobin concentrations between subjects and controls partially eliminates the problem of estimating specific individual doses. Carboxyhemoglobin concentrations in subjects experiencing sudden death due to ASHD were compared with those in subjects whose sudden death resulted from other causes or with the concentrations in living controls. For sudden death, the time between onset of the event and death is usually relatively brief. Since carboxyhemoglobin concentration is stable after death, the concentration determined at postmortem may be a good estimate of the concentration at the time of death.

TABLE 5-4 Relationship between Carboxyhemoglobin and Cause of Death

Study	Method	Results
Los Angeles County Chief Medical Examiner/Coroner[114]	20% sample, Los Angeles County Chief Medical Examiner/Coroner's case load: November 1950–June 1961, blood samples 2,207 cases; smoking questionnaire	1. High carboxyhemoglobin levels in cigarette smokers 2. Younger smokers had higher carboxyhemoglobin than older cigarette smokers 3. Greater the amount of smoking, higher the carboxyhemoglobin levels 4. Relationship between postmortem carboxyhemoglobin and ambient carbon monoxide among nonsmokers in most polluted areas 5. Cause of death had little or no relationship to the carboxyhemoglobin concentration at postmortem
Baltimore Office of the Chief Medical Examiner[244]	Comparison of carboxyhemoglobin concentrations in relation to cause of death. Age, race, sex for ASHD deaths by smoking history, detailed pathology, and length of survival; 2,366 deaths.	1. Carboxyhemoglobin higher for ASHD sudden-death subjects than in those of sudden death owing to other natural causes, but no difference in comparison with traumatic deaths 2. Among ASHD death subjects, carboxyhemoglobin higher in younger than older age group 3. Among smokers, carboxyhemoglobin higher in controls than ASHD death subjects. For nonsmokers, carboxyhemoglobin greater in ASHD sudden death than controls 4. No relation of ASHD death subjects' pathology, place of activity at onset, length of survival, and carboxyhemoglobin levels

The first such study took place in Los Angeles, where carboxyhemoglobin measurements were made at postmortem on a 20% sample of the subjects (2,207 deaths) from the Los Angeles County Chief Medical Examiner/Coroner's cases. Questionnaires about smoking habits were returned from the next-of-kin of 1,078 of the subjects (Table 5-4).[114] The cause of death was determined at the postmortem examination. Carboxyhemoglobin concentrations were higher in smokers than nonsmokers. In the most-polluted areas, there was a relation between the ambient carbon monoxide concentrations on a specific day and the mean postmortem carboxyhemoglobin concentrations. The causes of death showed little or no relation with the carboxyhemoglobin values taken at postmortem.

The Baltimore Study (Table 5-5)[244] measured the carboxyhemoglobin concentrations at postmortem for 2,366 subjects. These were compared by age, race, sex, and cause of death. The postmortem carboxyhemoglobin concentrations were higher in those who died suddenly owing to ASHD than in subjects whose sudden death resulted from other natural causes. This was true for all of the age group comparisons. Differences between the median concentrations of the groups were relatively small. No differences were noted between people who died from ASHD and people who died suddenly from traumatic causes such as accidents or homicides.

Smoking histories were obtained from a sample of the ASHD sudden-death subjects only, and carboxyhemoglobin concentrations were found to be substantially higher for smokers than nonsmokers. ASHD sudden-death subjects were compared with living controls. Living controls who were cigarette smokers had higher carboxyhemoglobin concentrations than ASHD sudden-death subjects who smoked cigarettes prior to death. For nonsmokers, the levels were higher for the ASHD sudden-death subjects than for living controls. However, when exsmokers were excluded, there were no differences in carboxyhemoglobin concentrations between individuals experiencing ASHD sudden death who were lifetime nonsmokers and similar living controls. There were also no differences in postmortem carboxyhemoglobin concentrations in the ASHD sudden-death subjects that were related to place of death, activity at onset, length of survival, and whether the deaths were witnessed or not.

A detailed pathology study was done of 120 ASHD sudden-death subjects whose postmortem carboxyhemoglobin concentrations were measured. Practically all of them had severe coronary arterial stenosis. There was no relation between the postmortem carboxyhemoglobin concentrations among the ASHD sudden-death subjects and the extent of coronary arterial stenosis, the presence of acute pathologic lesions such as a thrombosis or hemorrhage in the plaque, or recent myocardial infarction.

Neither of the two incidence studies, Los Angeles or Baltimore, revealed

TABLE 5-5 Median and Mean Carboxyhemoglobin Levels by Age and Cause of Death, Baltimore Sudden-Death Study[a]

	Age 25–34		Age 35–44		Age 45–54		Age 55–64	
Cause of Death	Median	Mean	Median	Mean	Median	Mean	Median	Mean
ASHD	1.5	2.0	1.6	2.6	1.9	2.6	0.7	1.6
Other natural	1.2	1.6	0.8	1.5	0.7	1.3	0.6	1.0
Auto accident	2.0	3.2	2.3	2.2	0.4	1.8	1.2	2.2
Other accidents	1.0	5.4	1.2	3.9	1.4	7.1	0.5	1.4
Homicide	2.4	2.9	2.0	2.7	2.9	3.7	1.2	2.1

[a] Reprinted with permission from Kuller *et al.*[244]

a relation between ambient carbon monoxide concentrations and the number of new ASHD cases. These two studies reported that there was a strong relation between carboxyhemoglobin concentrations and prior smoking history but little or no relation to the cause of sudden death.

Relation between Ambient Carbon Monoxide and Survival following a Myocardial Infarction

There has been only one study of the relation between ambient carbon monoxide concentration and case fatality following a myocardial infarction. Case-fatality rates were compared for patients admitted with a myocardial infarction to 35 Los Angeles hospitals during 1958. The carbon monoxide measurements were reported from monitoring stations operated by the Los Angeles County Air Pollution Control District.[244]

The hospitals were divided into those located in the low and those in the high carbon monoxide pollution areas. The low area was outside the 8 ppm isopleths for 1955. The majority of hospitals were located in the high area, and 2,484 patients with myocardial infarction were admitted in the high areas as compared to 596 admitted in the low areas. The areas more highly polluted with carbon monoxide had a greater case-fatality percentage than those less polluted during the weeks when the average carbon monoxide concentrations were in the highest quintile, 8.5 to 14.5 ppm weekly mean basin carbon monoxide concentrations. In 12 of the 13 weeks, the case-fatality percentage was greater in the higher carbon monoxide areas (Tables 5-6 and 5-7).

A further analysis of the data suggests that the differences in case fatality may not be a function of a change in ambient carbon monoxide

TABLE 5-6 Case-Fatality Percentage in Areas of High and Low Carbon Monoxide Pollution for the 13 Weeks in Which the Average Basin Weekly Mean Carbon Monoxide Concentrations Were Lowest[98]

Week of Year	Carbon Monoxide Average, ppm	High-Pollution Area Case Fatality (50)[a]	Low-Pollution Area Case Fatality (12)[a]
10	5.8	27	14
11	5.8	29	10
14	5.9	28	36
17	5.8	22	14
19	5.6	21	58
20	5.8	27	29
23	5.4	24	20
24	5.8	45	33
27	5.6	29	22
29	5.5	23	08
30	5.6	21	0
35	5.8	19	27
46	5.6	26	29
Median		26	27

[a]Estimated number of patients with myocardial infarction admitted to hospitals per week.

concentrations. During times of high pollution, the median percentile of case fatalities in the high-polluted area was 30%, and it was 20% in the low-polluted area. When the average basin pollution was low, the median percentile for case fatalities was 26% in the high-polluted areas and 27% in the low-polluted areas. Thus, as the mean basin carbon monoxide concentration increased, the case-fatality percentage increased slightly in the high-polluted areas but decreased in the low-polluted areas. It would be very unlikely that the case-fatality percentage would be inversely related to the carbon monoxide concentrations in the less-polluted areas. Practically all of the high mean basin carbon monoxide concentrations were reported in the winter, while the low concentrations occurred in the spring and summer. Because of the relatively small number of hospital admissions per week (12) in the low-pollution areas, there was a wide variation in mean case-fatality percentages (0–58%). These three factors suggest that the relation between ambient carbon monoxide concentrations and case-fatality percentages during high-pollution episodes may be related to a seasonal factor such as an influenza epidemic, changes in the number of hospital admissions in relation to the number of available beds, or other undetermined variables.

Seventy percent of all ASHD deaths take place outside of the hospital. The percentage of case fatalities among hospital admissions is an inadequate measure of the relation between environmental factors and the number of short-term fatalities following a heart attack. The percent of case fatalities within a hospital is a function of the incidence of heart attacks, their severity, and the rapidity of transfer to the hospital. When there is fast transportation to the hospital after heart attack, the number of case fatalities within the hospital might increase, while the overall percentage of case fatalities decreases.

Because of the significance for public health of a possible relation between case fatality and ambient carbon monoxide concentrations, particularly during periods of high pollution, replications of both the Baltimore and Los Angeles studies should be implemented. Such studies need to include the number of both in-hospital and out-of-hospital case fatalities, a description of the criteria both for diagnosing heart disease

TABLE 5-7 Case-Fatality Percentage in Areas of High and Low Carbon Monoxide Pollution for the 13 Weeks in Which the Average Basin Weekly Mean Carbon Monoxide Concentrations Were Highest[98]

Week of Year	Carbon Monoxide Average, ppm	High-Pollution Area Case Fatality (50)[a]	Low-Pollution Area Case Fatality (12)[a]
1	9.6	21	50
2	9.2	27	0
3	9.4	31	14
6	8.6	30	23
38	8.5	23	17
44	9.5	30	18
45	9.0	30	22
47	12.0	29	08
48	8.6	29	14
49	10.5	31	25
50	10.4	41	27
51	14.5	59	20
52	9.3	29	08
Median		30	20

[a] Estimated number of patients with myocardial infarction admitted to hospitals per week.

and for the demographic characteristics of the subjects, and effective methods for monitoring carbon monoxide.

Clinical/Experimental Studies of the Relation of Carbon Monoxide and Morbidity due to Heart Disease

Another approach to the study of the relation between exposure to carbon monoxide and the natural history of ASHD is to identify high-risk subjects and observe the effect of either natural or artificial carbon monoxide exposure. Such studies usually involve a technique for inducing symptoms in the subjects, such as exercise testing (Table 5-8).[6,13,14,15,16]

The first studies in 1971 compared subjects with angina pectoris before and after smoking nicotine-free cigarettes.[15] Ten male angina pectoris subjects smoked eight nicotine-free lettuce leaf cigarettes on two of four mornings. Following the smoking and/or nonsmoking mornings, the subjects exercised on a bicycle ergometer. Their carboxyhemoglobin concentrations rose to about 8.0% after smoking the cigarettes as compared to about 1% during the nonsmoking periods.

The duration of exercise prior to the onset of angina was reduced following cigarette smoking (Table 5-8). Chest pain also occurred at a lower systolic blood pressure and heart rate than for the nonsmoking mornings. Although prior to pain there was a wide difference in the duration of exercise, the duration was reduced for every man who smoked.

The next approach was to determine the effect of exposure to the high carbon monoxide concentrations on the Los Angeles freeway.[13] Ten patients with angina pectoris rode on the Los Angeles freeway for 90 min and then were brought to the laboratory for exercise testing. Exercise testing was done prior to the freeway trip, then immediately after it, and finally 2 hr later. Approximately 3 weeks later, the 10 subjects followed the same testing schedule except that during the 90-min freeway trip they breathed carbon-monoxide-free compressed air. Breathing carbon-monoxide-polluted ambient air, the mean carboxyhemoglobin increased to 5% during the freeway trip and decreased to 2.9% 2 hr later. Ischemic ST-T changes in the electrocardiogram occurred in 3 out of 10 subjects while breathing carbon-monoxide-polluted freeway air. There was a substantial decrease in the duration of exercise time before the onset of angina (both immediately after the trip and 2 hr later) in comparison to the exercise time prior to the freeway trip (Table 5-8). Breathing carbon-monoxide-free compressed air, there were no changes in exercise time before or after freeway travel. The subjects exercised longer after breathing the compressed air than after breathing carbon-monoxide-polluted freeway air. Systolic blood pressure and pulse were

TABLE 5-8 Clinical Studies of the Relation between Carbon Monoxide and Angina Pectoris

Study	Method	Sample Size	Carboxy-hemoglobin, %		Results	Reference
			Case	Control		
Carboxy-hemoglobin, nonnicotine cigarette (1971)	Evaluation of patients with angina following smoking nonnicotine cigarettes; exercise on bicycle ergometer	10 subjects; 2 smoked in morning, 2 nonsmoking	7.54 8.03	0.76 1.06	Reduction in exercise time to development of angina following smoking of cigarettes. No ECG changes	Aronow and Rokaw[15]
Freeway travel on angina pectoris (1972)	Comparison of breathing freeway air (high carbon monoxide) with carbon-monoxide-free compressed air; exercise bicycle	10 patients with angina pectoris, crossover	5.08 2.91 (2 hr later)	0.75 —	Reduction in mean exercise time, interval after freeway air, 3 patients with ST-T depression	Aronow *et al.*[13]

Carbon monoxide on exercise-induced angina pectoris (1973)	50 ppm carbon monoxide 2 hr for 2 mornings; carbon-monoxide-free compressed air 2 mornings; exercise bicycle	10 patients with angina pectoris, crossover, blind study	2.68	0.77	Breathing freeway air, reduction in length of time from exercise until onset of angina; no ECG changes	Aronow and Isbell[14]
Carbon monoxide exposure and onset and duration of angina pectoris (1973)	Carbon-monoxide-free compressed air, 50 ppm carbon monoxide, 100 ppm carbon monoxide	10 patients, crossover	2.8 (50—ppm) 4.51 (100—ppm)		Reduction in time to chest pain after exercise, no difference 50–100 ppm; S-T depression, ECG earlier	Anderson *et al.*[6]
Low-level carbon monoxide onset and duration, intermittent claudication (1973)	50 ppm carbon monoxide 2 hr compared to carbon-monoxide-free compressed air	10 men, crossover	2.97	0.90	Reduction in time to develop claudication; exercise on bicycle	Aronow *et al.*[16]

lower at the onset of angina after breathing the freeway air. No ECG differences were noted during the exercise testing, before the freeway trip, after it, or when breathing compressed air. These were not blind studies. The investigators and subjects both knew about the exposure to carbon monoxide. The subjects might even have suspected the possible occurrence of a harmful effect.

A double-blind study was done.[14] Ten subjects with angina pectoris were exposed either to carbon monoxide at 50 ppm for 2 hr on two mornings, or to compressed air for two mornings. They then exercised on a bicycle ergometer. The mean carboxyhemoglobin after 2 hr of the carbon monoxide exposure was 2.68% as compared to 0.77% with compressed air. Even at relatively low carboxyhemoglobin concentrations, the exercise time was reduced prior to the onset of angina pectoris. There was also a decrease in systolic blood pressure and heart rate at the time of onset of angina.

Anderson *et al.*[6] did a similar double-blind study in North Carolina. Ten subjects with angina pectoris were exposed to air with carbon monoxide at either 50 or 100 ppm for 4 hr. After exposure the subjects exercised on a Collins treadmill. The duration of exercise before the onset of chest pain was significantly shorter after exposure to carbon monoxide either at 100 ppm (mean carboxyhemoglobin, 4.5%) or at 50 ppm (mean carboxyhemoglobin, 2.8%). There was, however, no difference in exercise times after exposure to either 50 or 100 ppm of carbon monoxide. The S-T segment depression of the electrocardiogram generally appeared earlier and was deeper after breathing carbon monoxide. Other measures of cardiac function such as systolic time intervals, left ventricular ejection time, preejection period index, and preejection peak to ejection time ratio were within normal limits.

Subjects with intermittent claudication* have also been studied with this approach. In a double-blind study, they were exposed for 2 hr either to carbon monoxide at 50 ppm or to compressed air, and then they exercised.[16] Time until pain, in this case intermittent claudication rather than angina pectoris, was reduced after breathing carbon monoxide.

Only one experimental animal study of the effects of carbon monoxide on the natural history of heart disease has been reported. DeBias *et al.*[118] studied the effects of continuous exposure to 100 ppm carbon monoxide (115 mg/m^3) for 24 weeks in the cynomolgus monkey. They observed

*A complex of symptoms characterized by absence of pain or discomfort in a limb when at rest, the commencement of pain, tension, and weakness after walking is begun, intensification of the condition until walking becomes impossible, and the disappearance of the symptoms after a period of rest.

that carboxyhemoglobin chronically raised to an average of 12.4% produced polycythemia, and the hematocrit increased from 35 to 50%. All animals showed an increased P-wave amplitude and T-wave inversion in their electrocardiogram, suggestive of myocardial ischemia. Animals in which an experimental myocardial infarction was produced that were next exposed to carbon monoxide had more severe electrocardiographic changes than animals that breathed ambient air.

These studies of heart disease morbidity after carbon monoxide exposure have important implications. The first question that needs to be investigated is whether the experimental model in man, angina pectoris subjects exposed to low doses of carbon monoxide followed by exercise on a bicycle or treadmill, has relevance to the situation in a community. It has been suggested that a carboxyhemoglobin concentration as low as 2.5% has a deleterious health effect.[420,464] All cigarette smokers and about 10% of the nonsmokers in the United States frequently have carboxyhemoglobin concentrations higher than 2.5%. Thus a large percentage of the United States population may be potentially at risk. The National Health Survey Examination reported that there were 3,125,000 adults, aged 18 to 79, with definite coronary heart disease and another 2,410,000 who were suspect.[165] Many people who have severe coronary arterial stenosis without any apparent clinical disease would also be at high risk when the concentrations of carbon monoxide in the air are relatively low.

If the results of the clinical studies are applicable to this large population at risk, then a major public health problem exists. Taking the current results at face value suggests only that, when patients with angina are exposed to low carbon monoxide concentrations for short periods, they cannot exercise as long on a bicycle or treadmill before developing chest pain as those breathing compressed air. There is no evidence from these results that the exposure to carbon monoxide increases the frequency and severity of chest pain or the development of other complications or that it shortens life expectancy among patients with angina pectoris or other clinical manifestations of heart disease. We can only infer the existence of such a relationship.

Evidence for this association based on observations is equivocal. People with angina pectoris who smoke cigarettes can be expected to have higher carboxyhemoglobin concentrations and poorer prognosis than those who do not smoke cigarettes.[469] The harmful effect could be due both to carboxyhemoglobin and to other chemicals in cigarette smoke. A recent study has shown, on the other hand, that relatively few heart attacks occur while an individual is cigarette smoking and is thus exposed to higher carbon monoxide concentrations.[243] There is no evidence of an increased heart attack risk, including myocardial infarction and sudden death, while driving an automobile, and there are no data showing the occurrence of episodes of angina pectoris pain in relation to specific activities.

There is also no evidence suggesting a higher incidence, prevalence, or prognosis of heart disease among industrial workers who are exposed to high carbon monoxide concentrations.

And finally, there is little positive or negative evidence indicating that high ambient carbon monoxide concentrations in a community are associated with either the prevalence of angina pectoris or the natural history of heart disease. Furthermore, within communities, there is no evidence that there is a relation between the frequency of angina pectoris pain episodes and changes in the ambient carbon monoxide concentrations. Without such evidence, although the clinical experimental studies suggest important relationships between carbon monoxide and heart disease, ambient carbon monoxide cannot be implicated as a major causative factor of heart disease in a community.

Relation between Cigarette Smoking and Clinical Coronary Arterial Disease

The association between cigarette smoking and clinical coronary arterial disease is much closer for myocardial infarction and sudden death than it is for angina pectoris.[404] At present it is not completely clear why this is so, but it may be related to specific precipitating factors rather than to the underlying atherosclerosis. The association of cigarette smoking with clinical coronary arterial disease apparently depends on other risk factors, particularly a high-fat diet and increased serum cholesterol. This association is relatively weak in populations with low serum cholesterol content, but is apparently stronger in younger than in older people.[225]

Cigarette smoking appears to increase the risk of sudden death and myocardial infarction among subjects with preexisting angina pectoris.[469] On the other hand, the relation between smoking after a myocardial infarction and subsequent survival is less clear.[110,470] If cigarette smoking did not increase the mortality risk after a myocardial infarction, specifically among those who have survived for a month or so after the initial myocardial infarction, then the association between carbon monoxide and heart disease would be substantially weakened. A critical problem has been to try to separate the carbon monoxide effects of cigarette smoking from the effects of other harmful substances in cigarette smoke. After individuals free of clinical coronary disease cease smoking their risk of heart attacks is reduced. This reduction in risk takes place very soon after the cessation of cigarette smoking.[166]

The above observations suggest that a major effect of cigarette smoking may be as a precipitant of heart attack rather than in the development of the underlying atherosclerosis. There is a small amount of data suggesting that cigarette smokers have more extensive atherosclerosis than nonsmoking age-related controls. Such studies, however, do not have data adjusted for serum cholesterol levels or other risk factors. The effects of cigarette smoking on the incidence or clinical history of ASHD do not necessarily have to be due to carbon monoxide inhalation. Other factors in cigarette smoke, including nicotine, cyanide, or trace elements, may be important.

Studies relating carbon monoxide, smoking, and heart disease include 10 angina pectoris subjects who smoked cigarettes and who had blood pressure, heart rate, and expired carbon monoxide measurements taken before and after smoking high-, low-, and no-nicotine cigarettes.[11] After they smoked these cigarettes, their expired carbon monoxide concentrations increased with little difference among the three types of cigarettes. There was a significant increase in heart rate and systolic blood pressure after smoking the high- and low-nicotine cigarettes, but no effect after smoking nicotine-free cigarettes. Another study compared the effects of carbon monoxide and nicotine on cardiovascular dynamics in 8 men with angina pectoris.[10,12] Right and left heart catheterizations were done before and after cigarette smoking and then repeated after breathing of carbon monoxide at 150 ppm so that coronary sinus carbon monoxide would be similar to that after the smoking of three cigarettes. None of the subjects developed symptoms of angina pectoris either after smoking or after exposure to carbon monoxide. Smoking caused increases in the aortic systolic and diastolic blood pressure and in the heart rate.

These changes were not observed after breathing carbon monoxide. The left ventricular end-diastolic pressure increased after both smoking and carbon monoxide inhalation, the stroke index was reduced with both procedures, and the cardiac index did not change after smoking but declined after carbon monoxide inhalation. Smoking and carbon monoxide exposure also reduced the coronary sinus oxygen. Most of the effects disappeared or were substantially reduced within 30 min after the exposure. Therefore, after smoking, nicotine apparently increased the systolic and diastolic blood pressure and the heart rate, while carbon monoxide caused a negative inotropic effect, an increase in left ventricular end-diastolic pressure, and a decrease in the stroke index.

Wald *et al.*[463] have attempted to determine whether cigarette smokers with clinical coronary disease have higher carboxyhemoglobin concentrations than smokers without disease after adjusting for the amount of cigarette smoking. Volunteers (1,085) were recruited from several firms in Copenhagen, Denmark. They completed a questionnaire concerning their ASHD history. All who gave a positive history of heart disease were examined, and the history was validated both by examination and review of the medical records. Smoking histories were obtained from all of the subjects, and their carboxyhemoglobin was measured. A higher prevalence of ASHD was found with increased cigarette smoking. Men whose cigarette smoking was moderate to heavy and who had higher carboxyhemoglobin concentrations had a greater prevalence of ASHD. The relation persisted after adjusting for age, sex, duration of smoking history, serum cholesterol levels, and cigarette consumption. Although these results suggest a specific carboxyhemoglobin effect, they might also just be a measure of the inhalation of smoking products and not necessarily only a function of carbon monoxide inhalation.

The significance of carbon monoxide inhalation from smoking cigarettes is often dismissed when the effects of cigarette smoking on cardiovascular disease are evaluated. There is a good correlation with the amount of tar and nicotine in cigarettes, but not necessarily with the amount of carbon monoxide produced. A controversy exists about the effect on the increase in carboxyhemoglobin concentrations of smoking low- as compared to high-nicotine cigarettes. Several investigators have suggested that the carbon monoxide production of a cigarette be included on the package. A safe cigarette should be low in tar, nicotine, and carbon monoxide production. A cigarette that is low in carbon monoxide production has been made.

Cigarette smoking is the chief source of the high carboxyhemoglobin concentrations in the population. To effect major reductions in the mean carboxyhemoglobin in the population will require both a significant reduction in smoking and the modification of cigarettes to deliver lower doses of carbon monoxide.

The potentially harmful effects of low doses of carbon monoxide with respect to cardiovascular disease should provide further impetus to efforts to reduce cigarette smoking. People exposed to carbon monoxide from other sources, such as their occupations or concentrations in the ambient community air, may be at particularly high risk from cigarette smoking. The cigarette smoker exposed to high carbon monoxide concentrations in the ambient air will also have a greater carboxyhemoglobin concentration and a concomitant increased risk of heart attack.

The studies on cigarette smoking and cardiovascular disease are summarized in Table 5-9.

BEHAVIORAL EFFECTS

The experimental studies of carbon monoxide's effects on human behavior are reviewed under seven topical headings: vigilance, driving, reaction time, time discrimination and estimation, coordination and tracking, sensory processes, and complex intellectual behavior. Although carbon monoxide is probably the most widely studied of all toxic substances, our knowledge of its effect on behavior is limited.

Vigilance

Psychologists study vigilance by examining how well an individual performs when detecting small changes in his environment that take place at unpredictable intervals and so demand continuous attention.[283,284] Vigilance was first explicitly studied during World War II, when the British government became concerned about the performance of men who spent long hours searching for submarines or aircraft. This was a monotonous task and it was found that after a while the men would miss signals that they would not have missed at the start of their vigil.[285]

There have been a series of reports of carbon monoxide's effects on such a task. Groll-Knapp *et al.*[175] exposed "20 subjects of both sexes" to carbon monoxide at 0, 50, 100, or 150 ppm for a 2-hr period. Whether the subjects smoked is not stated. Carboxyhemoglobin concentrations were not measured directly, but rather it was estimated that by the end of exposure their values would have reached about 3% at 50 ppm, 5.4% at 100 ppm, and about 7.6% at 150 ppm. The subjects began an acoustic vigilance test 0.5 hr after exposure started. At this time the carboxyhemoglobin concentrations would have reached only about 1.5, 2, and 2.5%, respectively. Pairs of short tones, 1.1 s apart, were presented in a regular sequence. Over the 90-min test period, 200 of these pairs were transformed into signals by making the second tone "slightly weaker" than the first. The subjects reported their presence by pressing a button. Thus, signals occurred irregularly 2.2 times per min. Every subject was exposed to each of the four conditions once with the sequence systematically changed. The mean number of signals missed during the test was 26 at 0 ppm (control), 35 at 50 ppm, 40 at 100 ppm, and 44 at 150 ppm. Apparently even the smallest difference was significant. (A failure to replicate these findings is reported but not described in detail in a recent symposium paper from the same laboratory.[182])

Fodor and Winneke[141] carried out a second auditory signal study. They use a broad-band noise that lasted 0.36 s and was repeated at 2.0-s intervals. About three of the noises out of every hundred were made to

TABLE 5-9 Cigarette Smoking, Carbon Monoxide, and Cardiovascular Disease Studies Summarized

Study	Population	Methods	Results
Wald *et al.*[463]	1,085 volunteers, several firms, including tobacco companies	History of ASHD, validated by records and exam, smoking history and carboxyhemoglobin concentrations	Higher prevalence of ASHD with increased tobacco smoking, with increase in carboxyhemoglobin; slight effect on carboxyhemoglobin within smoking groups. At older ages, carboxyhemoglobin most powerful discriminator
Aronow *et al.*[11]	Comparison of nicotine and carbon monoxide effects; 10 men with angina; high-, low-, and no-nicotine cigarettes	(1) Significant increase in peak systolic and diastolic blood pressure from smoking high- or low-nicotine cigarettes. (2) Significant increase in heart rate with high- or low-nicotine cigarettes. (3) No change in heart rate or systolic blood pressure after smoking nicotine-free cigarettes	

Aronow *et al.*[12] (cigarette smoking and breathing carbon monoxide; cardiovascular hemodynamic in anginal patients)	8 men with angina pectoris	8 men with angina, right and left cardiac catheterization before, after smoking 1, 2, 3 cigarettes, replicate after 150 ppm—until coronary sinus carbon monoxide is same as after cigarette 3	None of the patients developed angina pectoris. Smoking, aortic systolic and diastolic blood pressure. Blood pressure, no change after breathing carbon monoxide. Heart rate increased with smoking, not with carbon monoxide. Left ventricular end-diastolic pressure increased after smoking, also after breathing carbon monoxide. Cardiac index unchanged after smoking, decreased with carbon monoxide. Stroke index decreased after both smoking and breathing carbon monoxide. Increase in coronary sinus oxygen after smoking or after carbon monoxide. Most of the hemodynamic changes reduced after 30 min.
Russell *et al.*[392] (effects of changing to low-tar, low-nicotine cigarettes	22 cigarette smokers	Comparison of increase in carboxyhemoglobin after smoking extra-strong cigarettes	Carboxyhemoglobin increase after smoking single strong cigarette was 1.45%, 1.09% for small brand, and 0.64% for extra-mild brand

be slightly less intense. These were the signals the subjects reported by pressing a button. Twelve subjects (male and female nonsmokers, 22–35 yr old) were tested at carbon monoxide concentrations of both 0 and 50 ppm, half with each order of presentation. Exposure lasted 80 min before the first 45-min vigilance test. A 5-min visual task followed the first vigilance test, after which there was a second 45-min vigilance test. Following another 5-min visual task, a third vigilance test was presented. The signals occurred randomly, with 44 of them occurring within any one 45-min period, or about 1/min. Carboxyhemoglobin concentrations were not determined directly but were predicted to be 2.3 and 3.1% at the beginning and end of the first vigilance test. The values for the second test were 3.1 and 3.7%, and 3.7 and 4.3% for the last test. Background carbon monoxide was neglected.

The percentage of signals missed is shown in Figure 5-19. Exposure to carbon monoxide caused the subjects to miss signals during the first vigilance test. This effect was not apparent during the next two vigilance test periods, and the interaction between carbon monoxide concentrations and time was statistically reliable. The data for the speed of response to the detected signals also showed a trend toward a maximum effect during the first period, but this time the interaction was not significant.

Fodor and Winneke[141] note that the "variation in performance after the 125th minute of exposure is hardly consistent with existing theoretical and experimental knowledge" and go on to speculate that "an organism exposed to carbon monoxide possesses the capability of compensation, which can, within certain limits, counterbalance a drop in performance." However, they recognize that "published data give little support to this *ex post facto* hypothesis, which urgently needs corroboration by replication experiments."

Winneke[491] reported another study of the effect of carbon monoxide on auditory vigilance in which the carbon monoxide data were collected along with data on methylene chloride, a compound suspected of exerting at least some of its behavioral effects by increasing carboxyhemoglobin concentrations. Eighteen subjects, nine male and nine female, were exposed to carbon monoxide at 0, 50, and 100 ppm on different occasions for almost 4 hr and then tested with the same auditory vigilance task used by Fodor and Winneke.[141] The results were completely negative, with differences not even in the expected direction (see Figure 5-20). Winneke[491] estimated that the subjects had an average carboxyhemoglobin concentration of about 9% after the 100 ppm exposure. At first glance the positive results with methylene chloride reinforce confidence in the reliability of the carbon monoxide data and are particularly welcome in an area where data are rare for substances that can serve as positive

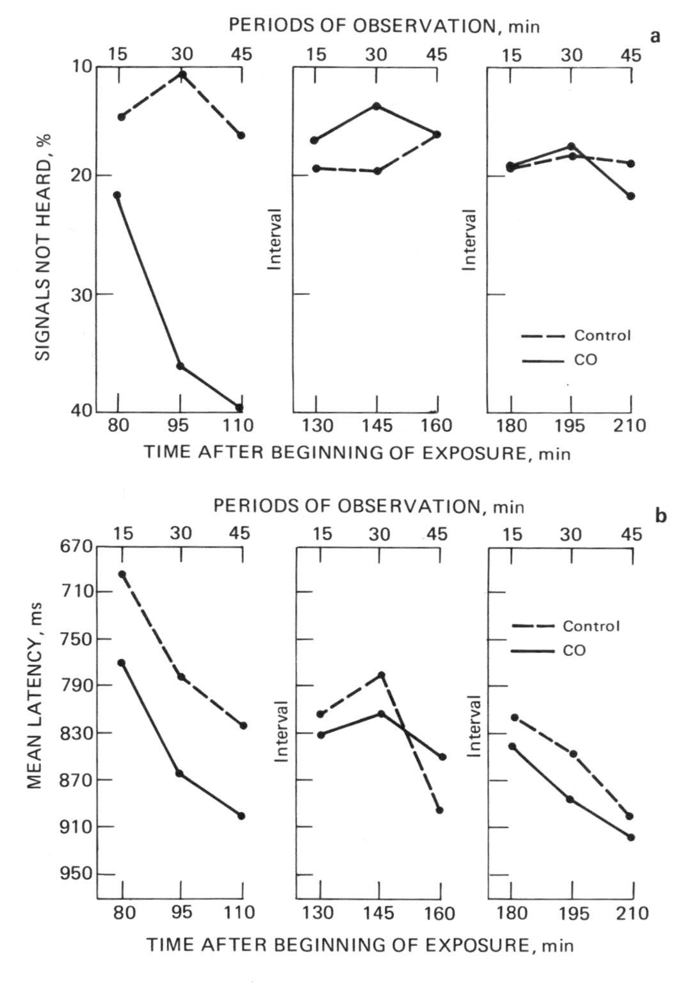

FIGURE 5-19 Effects of carbon monoxide on an acoustic vigilance test. (a) Number of signals missed after exposure to carbon monoxide at 0 and 50 ppm. (b) Latencies of responses to the detected signals. The panels show results from three successive 45-min vigilance tests. (Reprinted with permission from Fodor and Winneke.[141])

controls or reference substances.[198,250] However, this confidence is diminished by the inversion of the 300 ppm and 500 ppm curves and the irregular carbon monoxide control curve.

Horvath *et al.*,[199] O'Hanlon,[335] and O'Donnell *et al.*[334] studied carbon monoxide's effects on a visual vigilance task. A disk with a 1-in. diameter about 3 ft (0.9 m) from the subject was lit for 1 s every 3 s. The signal was a slightly brighter pulse. The subject pressed a button whenever he saw this brighter light. Before starting the test, each of 10 nonsmoking, "healthy male volunteers" 21–32 yr old were exposed via a mouthpiece for 1 hr to the same carbon monoxide concentration used during the test. The subjects were then brought to the experimental room where they were exposed further to carbon monoxide via another mouthpiece. Preceding the 1-hr main vigilance task, there was a short pretest called an "alerted" test, during which 10 out of 60 light pulses were the randomly interspersed, slightly brighter pulses that were the signal. These were presented at a signal rate of 3.3/min.

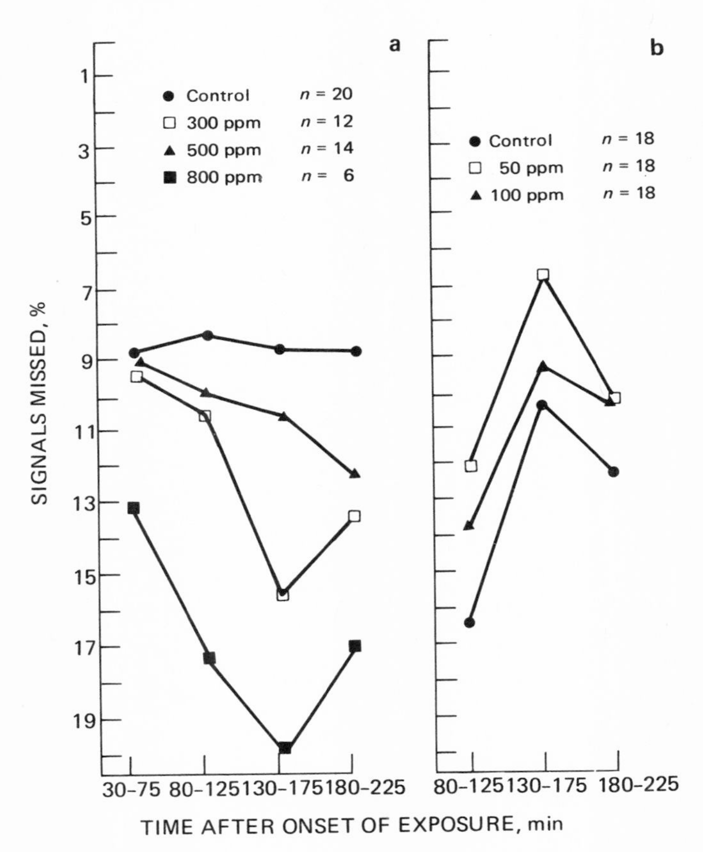

FIGURE 5-20 Vigilance performance after exposure to methylene chloride (a) and carbon monoxide (b). (Reprinted from Winneke.[491])

The 1-hr main vigilance task began after a 1-min rest. Twelve hundred light pulses were shown, 40 of which were the slightly brighter signals. These were distributed so that there were 10 among the 300 light pulses presented during each 15-min period. Therefore, the signal rate was 0.67/min. Since each subject was exposed at three different times to carbon monoxide concentrations of 0, 26, and 111 ppm, he served as his own control. The experiment was a single-blind study, with the subjects being unaware of the experimental treatment. The carboxyhemoglobin concentration of the controls (0 ppm) was the same, 0.8% after the initial-hour exposure as at the experiment's end. For the exposure to 26 ppm the carboxyhemoglobin concentration was 1.6% after the first hour and 2.3% at the end; and for the 111 ppm exposure the concentration was 4.2% after the first hour and 6.6% at the end. Performance on the pretest, when the signal rate was 3.3/min, was approximately the same for all three exposures. The carbon monoxide apparently had no effect. During the 1-hr vigilance test, subjects exposed to carbon monoxide at 111 ppm made significantly fewer correct signal identifications than did those same subjects exposed to either 0 or 26 ppm. The data are summarized in Figure 5-21. When the signal rate was 0.67/min, performace accuracy was reduced at a carboxyhemoglobin concentration of about 5%. However, during the pretest, when the signals appeared five times as frequently, there was no such effect. No measures of variability are given. The sudden increase in detection rate sometimes seen near the end of the hour may be an example of the end spurt often reported in vigilance work when subjects have been told when to finish their task.[285]

Beard and Grandstaff[35] examined the effect of carbon monoxide on a visual vigilance task also. In their double-blind experiment, the signal was a slightly shorter flash of light than that usually programmed to occur. The subject was seated in a small audiometric booth and faced a 3 in. (7.6 cm) electroluminescent panel 2 ft (0.6 m) from his eyes. Every 2 s this panel was lit. On most occasions the panel lit up for 0.5 s, but once in a while it remained lit for only 0.275 s; these were the signals to be reported by pressing a button. The subjects monitored the light flashes for four 30-min periods during the first day's session. The first of these periods was a control period. Carbon monoxide was then administered at concentrations of 0, 50, 175, or 250 ppm. A second 30-min vigilance task was begun after 10 min of exposure to carbon monoxide. This was followed by a 10-min rest period, after which there was a third 30-min vigilance test. The carbon monoxide was then shut off, and, after another 10-min break, the final 30-min vigilance test was given. During each of

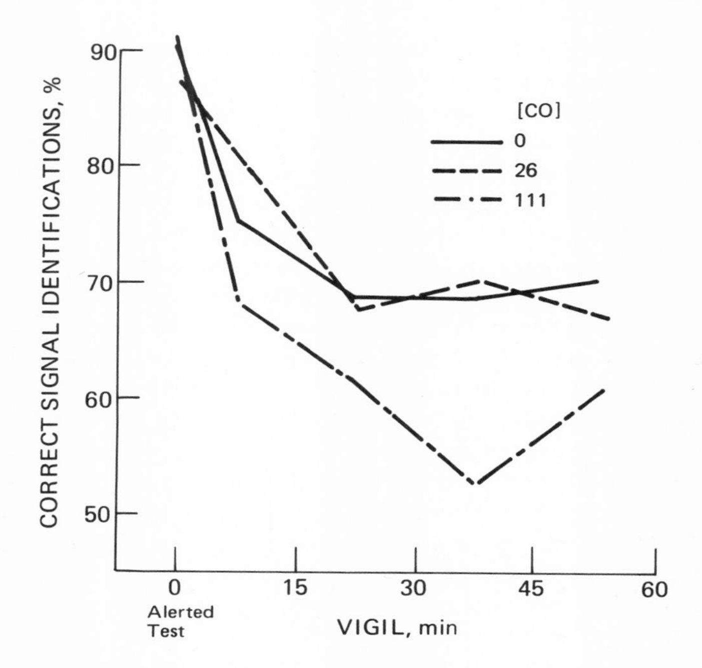

FIGURE 5-21 Vigilance performance after exposure to carbon monoxide. (Reprinted with permission from Horvath *et al.*[199])

the vigilance tests, 10 signals were randomly distributed among a total of 722 nonsignal light flashes, making the signal rate 0.33/min.

Nine nonsmoking male and female students, 18-25 yr old, were each tested in random order eight times, twice under each condition. Carboxyhemoglobin concentrations were estimated from alveolar breath samples taken both before testing and after the last vigilance test, 30 min after the cessation of the carbon monoxide exposure. The pretest concentrations were all below 1%. Post-test they averaged 1.8% for exposure to 50 ppm, 5.2% for 175 ppm, and 7.5% for 250 ppm. During the last three tests, about 73% of the signals were detected by subjects exposed only to room air. The detection rate was about 64% for exposure to both 50 and 175 ppm and 70% for exposure to 250 ppm. The decreases following exposure to 50 and 175 ppm, while small, were statistically significant at the 0.05 level (one-tailed test). The decrease following the 250 ppm exposure was not significant.

Further analysis of the data,[435] within a framework provided by signal-detection theory, showed that the significant changes in performance were related to both an increased degree of caution in making decisions and a decreased sensitivity (i.e., beta increased and *d* decreased). Interpretation of the peculiar results found with 250 ppm would be easier if levels high enough to produce pronounced effects had been included.

An experiment by Lewis *et al.*[259] should also be noted. These investigators examined the effects of traffic pollution on vigilance by exposing subjects to air pumped from the roadside of a busy street into the automobile that served as the experimental "room." Carbon monoxide levels were not measured during the testing but sometime after the test the experimenter determined that they were above 10 ppm about 40% of the time and above 30 ppm more than 4% of the time. The air was not analyzed for other noxious constituents. It is not clear how long subjects were in the immediate environment and therefore exposed to the flow of pollution before starting on these tasks. While seated in the test car they performed on an auditory vigilance task in which they were presented 0.5-s tones every 2 s. On nine occasions during a 45-min session, the tone was slightly shorter. Thus the tones occurred at the rate of about 0.2/min.

Subjects performed on the vigilance task twice, working on a number of other psychological tasks between runs. The 16 subjects used were 18–28 yr old. Order of testing was counterbalanced; one-half was first exposed to pure air, while the other half was first exposed to the polluted air. The subjects detected 73% of the signals while being given pure air from a metal cylinder but only 60% while breathing polluted air from the roadside. This difference was reported to be reliable but, as the authors realize, given the almost complete lack of knowledge of what the subjects were exposed to, one can conclude nothing about the specific effects of carbon monoxide. For instance, Fristedt and Akesson[147] found that workers in service stations located in multistory garages reported headaches and general fatigue at only slightly elevated lead and slightly elevated carboxyhemoglobin levels. They point out that exposure to neither substance was:

> of a magnitude such as to constitute a primary toxicological hazard. The combination of these in combination with other substances existing in automobile exhausts, such as acrolein and other aldehydes, alcohols, phenols, acetone, partially burned hydrocarbons, oxides of nitrogen, sulfur dioxide, and organic lead compounds, may pose a hazard to health, or a sanitation problem, manifesting itself in the form of discomfort. There never has been an exhaustive investigation of the biological reaction to chronic exposure to small quantities of a mixture of these substances.

Several investigators have looked for changes in physiological responses that may be related to vigilance. O'Donnell *et al.*[333] examined how overnight exposure to low carbon monoxide concentrations, with carboxyhemoglobin concentrations reaching 12.7%, affected sleep. They reported small and unreliable changes that were interpreted as a possible reduction in central nervous system activation. Their observations agreed with the findings of Xintaras *et al.*[496,497] on the evoked response in the rat and with Colmant's[102] data on disturbed sleep patterns in the rat. Xintaras *et al.*[496,497] found that effects on the visual-evoked response of the unrestrained, unanesthetized rat were similar for both carbon monoxide and pentobarbital, the classic reference standard for depression of reticular activity. They concluded that the changes induced by both substances resembled those recorded during the normal transition from wakefulness to sleep. In man it has also been possible to find carbon-monoxide-induced changes in the visual-evoked response. Carboxyhemoglobin concentrations of 20 to 28% were required,[201,425] which are much higher than those associated with vigilance changes in the studies previously cited. It is uncertain whether recent advances in the techniques for its measurement and analysis[370] will show an increase in this response's sensitivity to carbon monoxide.

In addition to studying behavioral vigilance, Groll-Knapp *et al.*[175] measured the slow-wave brain potentials presumed to correlate with anticipatory reactions to an oncoming signal. For their vigilance task, subjects were presented with a pair of brief tones. They had to respond if the second tone was weaker than the first. The researchers recorded for a 4-s period starting with the first pair of tones presented. They reported that, during a 90-min test period that started 30 min after exposure to a carbon monoxide concentration as low as 50 ppm, both the height reached by the anticipation wave after the first tone and the drop after the second tone were reliably reduced.

With vigilance as with other functions that will be discussed below, the interpretation of negative results with small amounts of carbon monoxide frequently cannot be made unambiguously. Before attacking the question of thresholds, investigators should demonstrate the sensitivity and specificity of their behavioral tests by employing doses high enough to produce measurable effects. They should also consider using previously studied drugs as reference substances.[198,250] Only after sensitivity and specificity have been established can one have confidence in negative findings with low concentrations.

Driving

In the earliest reported study (1937) of simulated automobile driving, Forbes *et al.*[142] exposed five subjects to enough carbon monoxide to produce carboxyhemoglobin concentrations as high as 30%. They reported very little effect on a series of reaction-time, coordination, and perceptual tasks presented within the context of a driving-skills test. The only control observations were made just before exposure, and there was no attempt to ascertain how much performance would have changed if room air alone had been administered. Consequently, their results cannot be interpreted and are only interesting historically.

McFarland and co-workers[296,298,299] have recently studied actual driving performance, focusing on two aspects of driving. One of these is the amount of visual information required by a driver. This was measured by asking the subject to maintain a constant speed while looking at the road as infrequently as possible. The subject wore a helmet with a shield that was placed in front of his eyes so that he was prevented from seeing the road. By depressing a foot switch he could briefly raise the shield. He was instructed to do this sufficiently often to keep his car within a 12-ft-wide marked lane on a deserted superhighway while maintaining a constant speed of either 30 or 50 mi/hr (48 or 80 km/hr) in different trials.[299] The number of steering wheel reversals was also monitored. Ten subjects were used in these experiments, with each acting as his own control. They were either exposed only to air or to enough carbon monoxide at a concentration of 700 ppm (via a mouthpiece) to produce a carboxyhemoglobin concentration level of 17%. McFarland reported that carboxyhemoglobin did not produce a differential effect on the frequency of steering wheel reversals. His conclusion concerning visual interruption data is less clear. From a significant interaction term in an analysis of variance, McFarland concluded that those subjects exposed to carbon monoxide while driving at the higher speed required more roadway viewing than those exposed only to air.

Ray and Rockwell[369] examined the driving performance of three subjects with carboxyhemoglobin concentrations of 10–20% from carbon monoxide exposure. In this study the subject rode in a car yoked by a taut wire to a second car driven ahead of it. Information concerning relative velocity and the distance between the two vehicles was transmitted via the wire. In some experiments the subject drove, while in others he was a passenger. The authors only reported data for an experiment in which the subject attempted to detect slight changes in the relative velocity of the two vehicles when the lead vehicle was 200 ft (61 m) ahead of the car in which he rode as a passenger. The time required to respond to a velocity change of 2.5 mi/hr (4 km/hr) was approximately 1.3 s for control conditions, 3.3 s when the subjects had a carboxyhemoglobin concentration of about 10%, and 3.8 s when it was about 20%. These differences were statistically significant for a group of three subjects, each of whom had been exposed to all three conditions. In view of the small sample size, the authors wisely considered these results as only exploratory.

Weir and Rockwell[474] have made extensive observations of driving behavior. This as yet unpublished research also utilized yoking one car to another by a thin wire. Gas pedal positions, brake pedal applications, steering wheel reversals, actual velocity, and separation of the lead and following vehicles were recorded. The results were negative for carboxyhemoglobin concentrations of 7 and 12%. These were the concentrations studied with the largest numbers of subjects and the tightest experimental designs.

Another study of carbon monoxide's effects on driving performance used "a standard driving simulator,"[495] a device that simulated a 10-min drive through traffic, giving the subjects a brake pedal, an accelerator, and a steering wheel with which to react to various realistic driving conditions shown on a film. Forty-four adult volunteers were used. Half of these were randomly allocated to a group receiving only air and half to a group receiving enough carbon monoxide to produce carboxyhemoglobin concentrations about 3.4% higher than they had when tested before exposure. Half of the experimental group were smokers, and their carboxyhemoglobin concentrations just before the second driving simulator test averaged about 7.0% (up from about 4.4%), whereas the nonsmokers averaged 5.6% (up from about 1.3%). An 80-ml dose of pure carbon monoxide was introduced into the breathing system that was also used to make carboxyhemoglobin determinations. The subjects breathed this as a carbon monoxide concentration of 2%.* The carbon monoxide was not found to affect the overall performance on the driving simulator.

The experimenters divided the various individual activities that had been scored during the simulation task into two categories: "emergency actions" and "careful driving habits." No difference could be detected in the way the experimental and control subjects reacted on the category "emergency actions." With the "careful driving habits," the group working under the influence of carbon monoxide showed more deteriora-

*Administering carbon monoxide at a high concentration for a short time may produce greater effects on performance than the customary method of exposing subjects to low concentrations for longer times. While the carboxyhemoglobin concentration is usually thought to be the most important physiological stimulus for determining carbon monoxide's effects, there is some evidence that the rate of saturation is also important (see p. 165). Plevova and Frantik[354] recently (1974) demonstrated that exposing rats to 700 ppm for 30 min produced a greater decrease in the length of a forced run on a treadmill than did exposure to 200 ppm for 24 hr, even though both exposures brought the average carboxyhemoglobin concentration at the start of the experimental task to approximately 20%.

tion than the control group. The change is only marginally statistically reliable, however. The relevant data shown in the researchers' table have a chi-square of 5.84, which with 2 degrees of freedom yields a probability between 0.10 and 0.05. Consequently, this finding can only be regarded as suggestive support for their conclusion that a 3.4% increase in carboxyhemoglobin is sufficient to prejudice safe driving.[474]

Rummo and Sarlanis[388] also used a driving simulator to examine the effects of approximately 7% carboxyhemoglobin concentration. Carbon monoxide at 800 ppm was administered for 20 min prior to a 2-hr simulated drive. By also dosing the subjects in the same manner with air before one of the two experimental sessions, they kept the subjects unaware of when they received the carbon monoxide. The experimenter knew, however, which gas treatment was in effect. The seven subjects (six nonsmokers, one smoker) tried to maintain a specified distance between the automobile that they were "driving" and one that appeared to be in the same lane ahead of them. They were instructed to respond as quickly as possible to any changes in this separation. Both increases and decreases in the separation were programmed to occur at random intervals 10 times every 0.5 hr. Steering wheel reversals and braking responses to a red warning light that appeared on the dashboard for a few seconds eight times every 0.5 hr were also recorded. An increase in the response time to changes in the speed of the car ahead was associated with the carbon monoxide treatment. The mean response time increased from 7.8 s to 9.6 s, a statistically significant effect. It was also observed that fewer steering wheel reversals were made after carbon monoxide administration. The lone smoker showed the opposite effect. The change was statistically significant only when that subject's data were not included in the calculations. Several other measures—reaction time to the dashboard light, how much space the subjects kept between cars, and ability to maintain lateral position on the simulated road—showed no reliable changes. The authors pointed out one factor that may have helped them observe a reliable decrease in the rapidity with which changes in the distance between the two cars were detected. The task "was designed as a vigilance task that closely simulated a real-life situation of prolonged driving on a little-traveled road under twilight conditions. Nearly all the subjects remarked that the driving was realistic but boring."[388]

The paucity of published research on driving and carbon monoxide is somewhat puzzling, given the long-standing interest in the question by strong political and economic groups. Perhaps one impediment has been the relative difficulty in devising both suitable and safe experimental preparations. In light of the strong preference on the part of some that experimenters work with the precise behavior at issue, rather than with laboratory analogues,[76,77] the field is probably awaiting the development of adequate field methods (see, for example, Mourant and Rockwell[320]).

Reaction Time

Reports on the effects of carbon monoxide on reaction time have been conflicting. In the last few years several well-controlled studies were completely negative, whereas several other equally well-controlled studies were positive.

Negative studies were reported by Wright *et al.*,[495] Stewart *et al.*,[424] Fodor and Winneke,[141] Winneke,[491] and Rummo and Sarlanis.[388] The study by Rummo and Sarlanis is described above.

In another driving simulator study also described above, Wright *et al.*[495] had subjects release the accelerator pedal and depress the brake pedal as quickly as possible when a red light was lit by the experimenter. No reliable difference was found in performance on this task at carboxyhemoglobin concentrations about 3.4% higher than the subjects (both smokers and nonsmokers) had at the onset of the experiment.

Stewart *et al.*[424] used the American Automobile Association driving simulator in a reaction-time study. The subject was presented one of three stimuli, each of which signaled a different response: turning the steering wheel left or right or removing the foot from the accelerator and depressing the brake pedal. Carboxyhemoglobin concentrations as high as about 16% did not change the reaction time.

During their study of vigilance, Fodor and Winneke[141] carried out three different reaction-time tests: one in which the subject kept his finger directly above a response button, one in which the subject had to move 25 cm from where his finger rested between trials to the button, and one discriminative reaction-time test in which five buttons in a circle were paired with five different signals. No significant differences were found for any of these at carboxyhemoglobin concentrations estimated to be 2.3 and 5.3%. The same procedures also produced negative results in a study by Winneke[491] in which carboxyhemoglobin concentrations were an estimated 10%.

Ramsey[363,365] reported two studies in which longer reaction times with carbon monoxide were observed. In one study,[363] 60 subjects were exposed via a face mask to 300 ppm for 45 min, while 20 controls breathed only air from a tank through the mask. The mean carboxyhemoglobin concentration of the exposed group reached about 4.5%. The reaction-time test is described by the author only as "reaction-time to a visual stimulus." Subjects exposed to carbon monoxide showed a very small but statistically significant increase in reaction time. In this experiment, the 60 experimental subjects were of three different types: One subgroup of 20 subjects were patients with "mild anemia," a second 20 subjects suffered from emphysema, and the remaining 20 subjects were in normal health. There was no difference in the changes induced by carbon monoxide in these subgroups. In a second experiment, Ramsey[365] exposed 20 healthy male subjects to 650 ppm for 45 min and another 20 subjects to 950 ppm for the same time. The carboxyhemoglobin concentrations increased by 7.6% in the first case and 11.2% in the second. Twenty other subjects received only pure air. A discriminative reaction-time task was used in which the subject had to respond to various colored lights on different buttons. The increases in reaction time shown by the two experimental groups were both statistically significant.

The interpretation of many of the reaction-time studies is difficult because the experimental procedures have not been fully described. Even though stimulus intensity, intertrial interval, and the precise response required have all long been known to be important in reaction-time work, they are frequently unspecified.[436,437]

Time Discrimination and Estimation

In 1967, Beard and Wertheim[37] published an account of an experiment in which they examined the effects of carbon monoxide on the ability of human subjects to discriminate between the lengths of two tones. In each case the first tone, which was 1 s long, served as the standard. The second tone was presented 0.5 s later and varied in duration from 0.675 to 1.325 s. The subject's task was to report whether the second tone was shorter, longer, or the same length as the first tone. He was given three buttons with which to report his judgment. The pairs were presented in sets of 50, with approximately 7.5 s elapsing between the initiation of pairs. Within each half of this set, one-third of the comparison stimuli were identical to the standard, one-third were longer, and one-third were shorter. These were scrambled in a nonsystematic sequence. It took 6 to 7 min to complete the 50 trials. The subject then had about 13 min during which he could either read, rest, or watch television while remaining in the small audiometric booth used as an exposure chamber. The next set of 50 trials was signaled by a warning light and a tone. This cycle of working approximately one-third of the time and resting the other two-thirds was repeated three times per hour for 4 hr, so the subject made a total of 600 judgments during each experimental session.

Eighteen university students were each exposed three times to each of the following carbon monoxide concentrations: 0, 50, 100, 175, and 250 ppm.

The subjects were unaware of the concentrations being used, but the experimenter knew. The experimenters presented their results in terms of "mean percent correct responses." Their analysis of greatest interest concerns the amount of carbon monoxide exposure necessary to produce a marked performance decrement. Under control conditions, subjects averaged about 78% correct at 0 ppm (see Beard and Wertheim,[37] Figure 1), with a standard deviation of approximately 5 percentage points. The authors presented data on how long an exposure to various carbon monoxide levels was needed to produce a decrement in performance more than two standard deviations in magnitude (Figure 5-22). A decrease from 78% correct to approximately 68% incorrect was produced by 50 ppm carbon monoxide in about 90 min, by 100 ppm in about 50 min, by 175 ppm in about 32 min, and by 250 ppm in 23 min. Note that the subject had been in the booth and working for 30 min before the exposure began. Carboxyhemoglobin levels were not given, so they can only be estimated. According to Coburn *et al.*,[91] these levels would have ranged from approximately 2.5 to 4%.

Beard and Wertheim[37] also plotted correct responses versus carbon monoxide concentration in the booth. They found a linear function, with

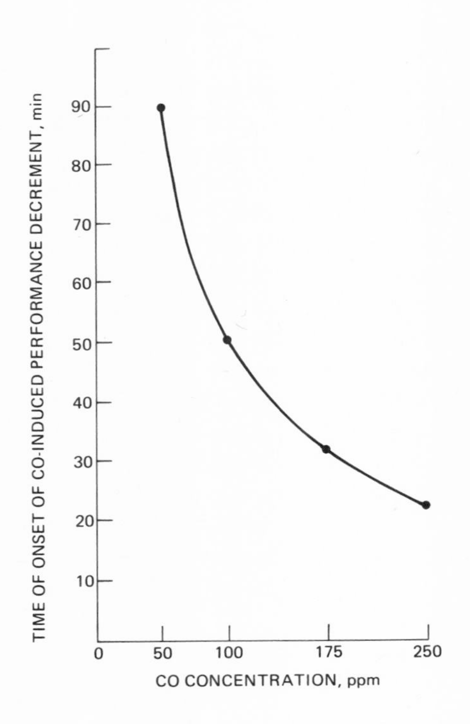

FIGURE 5-22 Effects of carbon monoxide on time discrimination. Time after initial exposure to each dose at which the mean percent correct responses fell below two standard deviations of the mean performance without carbon monoxide. (Reprinted with permission from Beard and Wertheim.[37])

subjects exposed to 250 ppm scoring fewer than 30% correct, whereas control subjects scored nearly 80% correct, these figures being derived from a 2-hr carbon monoxide exposure starting 1 hr after the subject had been placed in the booth to start the task and scoring started 0.5 hr after exposure had begun. Thus, they refer to carboxyhemoglobin concentrations estimated to average 2, 4, 6, and 9% for the 50, 100, 175, and 250 ppm exposures, respectively.

Although this experiment, which showed deleterious effects from rather short-duration exposures to low carbon monoxide concentration, was reported a decade ago, only three groups have attempted to replicate its findings. In each case, the replication has been less than satisfactory.

In 1970 Beard and Grandstaff[36] reported that "the earlier work with tone duration had been confirmed in 7 additional subjects." No new data were presented in their paper, however. In 1971 O'Donnell *et al.*[333] studied the effects of overnight carbon monoxide exposure on a temporal discrimination task patterned after that used by Beard and Wertheim.[37] A 1-s standard tone was delivered 0.5 s before a tone that varied in length between 0.675 and 1.325 s. Neither the duration of the task nor the total number of tone pairs was reported. This task was given twice during a 1.5-hr battery of tests that included two other tests of time estimation, two tracking and monitoring tasks, measures of critical flicker frequency, and mental arithmetic. All tests were done in the morning after the subject had spent approximately 7.5 hr in a special chamber exposed to the carbon monoxide concentration level at which he was being tested. This was a double-blind study in which neither the subjects nor the experimenters knew which exposure was being studied. There were four subjects, each of whom produced data on three out of nine nights spent in the experimental chamber. On the first four nights the subjects adapted to the experimental procedures. On the fifth night they were exposed to carbon monoxide at either 75 ppm or 150 ppm. On the sixth and eighth nights they went through all the experimental procedures but were not exposed. Data for these nights were not reported. On the seventh night they were exposed to the concentration alternate to that used on the fifth night, and on the ninth night, to 0 concentration. Because the subjects went through the experimental conditions at either 75, 150, 0 or 150, 75, 0 ppm, only the two higher carbon monoxide concentrations can be compared statistically. The carboxyhemoglobin concentrations were 5.9% for the 75-ppm exposure and 12.7% for the 150-ppm exposure. No difference was found between the subjects' temporal discrimination scores for the two concentrations; the means were 6.13 ± 1.53 at 5.9% carboxyhemoglobin and 6.50 ± 1.03 at 12.7%.

The third experiment on carbon monoxide's effects on auditory duration discrimination was carried out by Stewart and co-workers.[422,423] The subjects were exposed to carbon monoxide at different concentrations for varying lengths of time. This was also a double-blind study. Four different time-estimation methods were investigated. One was an auditory time-discrimination task modeled after the one used by Beard and Wertheim.[37] Stewart *et al.*[422,423] examined performance on this task under three different conditions: Twenty-four subjects were tested while seated along one side of a table in an experimental room measuring 20 × 20 × 8 ft (6 × 6 × 2.4 m); five subjects were tested one at a time in the experimental room; and nine subjects were tested in an audiometric booth placed inside the larger experimental room. Seventy-five pairs of tones were presented in sets of twenty-five with 30-s pauses between the sets. The time between the tone pairs is not reported. However, since the model was Beard and Wertheim's experiment, it is assumed that similarly the pairs were 7.5 s apart. According to the authors, the task took approximately 15 min. Other tasks probably took another 10 min to carry out. Since the entire task group was usually carried out once per hour during the carbon monoxide exposure, the subjects were free to interact with one another for at least part of every hour. This situation was quite different from the one in which Beard and Wertheim's subjects found themselves alone in the small booth working on their single task approximately one-third of the time and resting the remaining time.

Stewart *et al.*[422,423] presented their results in two ways. In one, they combined all the preexposure data with the data collected after exposure to the control concentration, described as air containing carbon monoxide at less than 2 ppm. They then contrasted the score at this "base line" concentration with the mean score on the tone-discrimination task at carboxyhemoglobin concentrations as high as 20%, which were reached after exposure to various carbon monoxide concentrations. With this approach there was no apparent effect. This method of presentation does not take into consideration that the subjects were being used as their own controls by producing both a pretest and a post-test for each carbon monoxide exposure. Therefore the experimenters could more precisely characterize each subject's reaction to the different exposures. Stewart *et al.*[422,423] also presented this type of analysis, not presenting means but presenting the results of t tests. The changes were not statistically significant for the two measurement conditions in which the subjects were not tested in the small booth. The t test was significant for the nine subjects who were tested in the booth. These subjects had a 2.9% mean decrement in the number of correct responses. With 8 degrees of freedom, the t value of 2.75 was significant at the 0.05 level. The authors reported that this decrement was associated with a 9.7% mean carboxyhemoglobin concentration. Beard and Wertheim's[37] subjects showed a much greater reduction in performance at much lower carboxyhemoglobin concentrations. However, the results may not be comparable because their experiment was not exactly duplicated.

Two other methods of studying carbon monoxide's effects on time perception have been used. In one, the subject was requested to respond at specified regular intervals. Beard and Grandstaff[36] studied the subject's ability to estimate the passage of either 10 or 30 s in this way. They reported that exposure to as little as 50 ppm for 64 min impaired the judgment of a 30-s period. Estimating a 10-s period, however, was not modified by longer exposures to as much as 250 ppm. Not enough experimental detail was reported to evaluate the data, and carboxyhemoglobin concentrations were not determined directly.

O'Donnell *et al.*[334] and Mikulka *et al.*[313] studied the effects of low carbon monoxide concentrations on the ability of men to space button presses 10 s apart. The subjects received no feedback concerning their accuracy. Data are reported for nine subjects, each exposed three times to 0, 50, or 125 ppm for a 3-hr period. The carboxyhemoglobin concentrations at the end of the exposures were 1.0, 3.0, and 6.6%, respectively. Time estimates were made for 3 min every 0.5 hr. This task was part of a 15-min-long battery that included some tests of tracking and ataxia. It was a double-blind study in that both the experimenters and the subjects were not informed about the exposures used. The mean time estimates were found to be higher under the influence of carbon monoxide.

The experimenters carefully counterbalanced exposure to the three conditions and did analyses of variance for each of the six time periods. Only one, at 135 to 150 min after the beginning of exposure, yielded a reliable difference. There was also no apparent trend over time toward greater differences between the control and experimental conditions, which would be expected if the carboxyhemoglobin concentrations were rising with carbon monoxide exposure. This, plus the greater difference found with exposure to 50 ppm than to 125 ppm, led the investigators to conclude that they had not demonstrated consistent carbon monoxide effects on this type of performance.

O'Donnell *et al.*[333] studied the discrimination of 1-s tones after overnight exposures and also studied discrimination of 10-s intervals. Because they did not counterbalance 0 exposure, which was always last, only their results with exposure to 75 and 150 ppm can be compared. These were completely negative.

Stewart and co-workers[422,423] also investigated the effect of carbon monoxide on the discrimination of 10-s time intervals. Their general technique is described above. In the 10-s time estimations, similar to those for 30-s intervals, the subject held down a push button for what he thought was the appropriate time. The mean of two such judgments was taken every 1 hr during exposure. The results for both the 10- and 30-s tests were negative.

Stewart and his associates[422-424] have examined carbon monoxide's effects when the subject had to reproduce an interval of time by depressing a button for the same length of time as the just-presented auditory or visual signal. The stimuli lasted either 1, 3, or 5 s and were presented in random order. Three stimuli of each duration constituted a test. The test lasted approximately 7 min. In one experiment,[424] the subjects were tested immediately after entering the experimental chamber after either 4- or 7-hr exposure to carbon monoxide, or, if the exposure was for less than 4 hr, during the last 0.5 hr of the exposure. There were no apparent effects at carboxyhemoglobin concentrations as high as 25%. In a later experiment in which carboxyhemoglobin concentrations rose to 20%, the results were only slightly less negative.[422,423] The earlier study had been done with the subjects working in one large room. Part of the later experiment was carried out under more controlled conditions. The subject worked either alone in a large room or in a small audiometric booth. This reduced the influence of factors extraneous to the study. The data were not presented in the published paper. The results of the *t* tests were cited that led the researchers to reject the possibility of a carbon monoxide effect. Both experiments were conducted under double-blind conditions.

Although in Beard and Wertheim's[37] original experiment the researchers knew which exposure a subject received (it was not a double-blind study and no direct measurements of the carboxyhemoglobin concentrations were made), this experiment's demonstration of the effects of very low carbon monoxide concentrations on time discrimination and estimation of behavior has not been refuted. Neither has it been confirmed or repeated by other workers. However, since either very small or no effects on other timing-behavior tests have been reported with carbon monoxide doses larger than those used by Beard and Wertheim, the effect they found may have been due to aspects of their task other than those associated with time perception. Judging from the frequent positive results reported by researchers studying vigilance (see above), it is possible that Beard and Wertheim's use of isolated subjects who were given a prolonged exposure to the tone-duration discrimination task was crucial to their positive findings. Both the research groups at Marquette University[422-424] and at Wright–Patterson Air Field[333,334] minimized boredom in their timing-behavior studies. For instance, O'Donnell and co-workers[334] reported that, "following 90 min of exposure, the subject was allowed to walk around and stretch in order to reduce the possibility of fatigue and boredom from being in a constant position for 3 hr." On the other hand, the Stanford researchers[35-37] invariably minimized external influences and ran their experiments for longer periods of time (cf. Nielsen[329]), thereby turning them into vigilance experiments. Thus, both groups may be correct in what they are reporting, with external stimulation being the variable that determines sensitivity to the effects of carbon monoxide.

Coordination and Tracking

The conclusions of most of the coordination, steadiness, dexterity, and tracking tests examined were negative. Stewart *et al.*[424] used several tests of manual dexterity in which, for example, subjects had to pick up small pins, place them in little holes, and then put collars over the pins. No effects were found, even at carboxyhemoglobin concentrations of about 15%. Changes in manual dexterity were seen in a separate series of observations on two subjects when concentrations of almost 30% were reached. Wright *et al.*[495] also found no effects on hand steadiness at carboxyhemoglobin concentrations of 7%. Fodor and Winneke[141] also reported negative results on a series of coordination tasks performed at carboxyhemoglobin concentrations estimated to be about 5% (from exposure to carbon monoxide at 50 ppm for about 4.5 hr). Winneke[491] also reported no change in similar task performance after doubling the exposure level to 100 ppm for about the same length of time.

Bender *et al.*[41,42] reported some positive results with a coordination task on which 42 students were tested. Each subject was exposed to both 0 and 100 ppm. The experimenter but not the subjects was informed which exposure was being studied. At carboxyhemoglobin concentrations estimated to be about 7% (2.5 hr exposure to 100 ppm), small but significant decrements were observed in how skillfully the subjects worked on the Purdue peg board, on which pegs, bushings, and rings had to be assembled.

In 1929 Dorcus and Weigand[126] reported that steadiness was the only characteristic tested that showed any apparent effect after 5-hr exposures to exhaust gas containing carbon monoxide at up to 400 ppm (cf. Sayers *et al.*[393]). More recently (1974), in a study of toll booth operators, Johnson *et al.*[214] found that a test of eye–hand coordination correlated significantly with the increase in carboxyhemoglobin concentration produced by exposure to automobile exhaust.

O'Donnell *et al.*[334] and Mikulka *et al.*[313] studied performance on a tracking task in which a needle had to be kept within prescribed limits on a dial for 1 min. While carrying out this task the subject also had to monitor three other dials, which occasionally showed deflections requiring adjustment. In another version of this task, an additional job was given. The subject monitored three lights that flashed occasionally and reported at the end of the 1-min tracking trial the total number of times each light had flashed. At carboxyhemoglobin concentrations of 12.7%, no effects were observed on either task. It is possible that these tasks would have been more sensitive indicators if they had been carried out for longer than 1 min. In psychopharmacology studies it has been found that prolonged task performances are more likely to be sensitive to change by chemical agents.[251,475]

Both O'Donnell *et al.*[334] and Mikulka *et al.*[313] studied a tracking task in which subjects reacted with corrective movements to needle deflections, which became more difficult to compensate with time. The length of time the subject could keep up with this increasingly difficult task was measured. Nine nonsmoking male students were each exposed for three periods of 3-hr duration to 0, 50, and 125 ppm. At the end of 3 hr, the carboxyhemoglobin concentrations were 1.0, 3.0, and 6.5%, respectively. Although there was a significant decrement in performance about halfway through the exposure, when the authors examined the individual curves they concluded that the differences were not statistically reliable. Hanks,[186] using the same tracking task, reported no effects at up to 100 ppm for 4.5 hr. His report contains no data, however.

O'Donnell *et al.*[334] gave their subjects the Pensacola Ataxia Battery when they had completed the tracking task. At carboxyhemoglobin concentrations up to 6.6%, no effects were observed on such measures of ataxia as standing on one leg or walking a straight line both with closed eyes.

Sensory Processes

Vision is the only sensory function that has received much attention. The early literature abounds with case histories of the sequelae of both acute and chronic carbon monoxide poisoning,[488] but there have been surprisingly few experimental studies. During World War II, McFarland and co-workers[184,297,300] studied brightness discrimination. They made very careful measurements with a small group of well-trained subjects, thereby minimizing the variability that usually makes the detection of small differences difficult. Figure 5-23 shows data from an experiment that compared a simulated altitude of 4,694 m (15,400 ft) with various carboxyhemoglobin levels produced by graded doses of carbon monoxide. Also shown are the effects of administration of pure oxygen

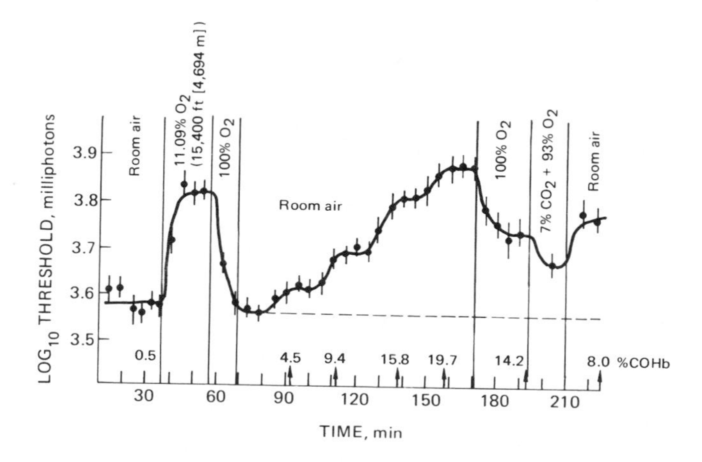

FIGURE 5-23 The effect of progressive increases in carboxyhemoglobin percentage in blood on brightness discrimination thresholds. Each point represents the mean of 10 measurements, and the vertical bars represent plus and minus one standard deviation. Carbon monoxide was administered at times indicated by the horizontal lines between arrowheads. (Reprinted with permission from Halperin *et al.*[184])

and a combination of 7% oxygen and 93% carbon dioxide (carbogen). Carbon monoxide was given via a face mask. It is clear that even slight increases in carboxyhemoglobin level, to about 4.5%, increased the measured visual threshold, indicating that the subject's ability to distinguish an increase in the intensity of a dimly lit field was diminished. In 1970 Beard and Grandstaff[36] reported that Wertheim found statistically significant increases in the brightness thresholds of four young adults and a decrease in vernier visual acuity at a carboxyhemoglobin concentration of 3% produced by carbon monoxide exposure.

Conversely, Ramsey[365] failed to detect any changes in brightness discrimination in a study of 60 young adults, 20 of whom were exposed to enough carbon monoxide to produce a carboxyhemoglobin concentration of about 8%, 20 to enough to produce a concentration of about 12%, and 20 controls who received only air. Using carbon monoxide concentrations that produced carboxyhemoglobin percentages approximately 3.4% above those at the beginning of exposure, Wright *et al.*[495] did not find any effect on several measures of visual function: night vision, glare vision, glare recovery, and depth perception. A possible change in the speed of dark adaptation was studied in 1973 by McFarland and co-workers[296,298] and earlier in 1944 by Abramson and Heyman,[1] with negative results in both experiments. McFarland's study used carboxyhemoglobin concentrations as high as 17%. However, none of these negative studies presented evidence of experimental control comparable to that shown in the earlier work by McFarland and his colleagues.[184,297,300]

Several investigators have examined the effects of low carbon monoxide concentrations on critical flicker fusion frequency (CFF), the frequency above which an intermittent light appears not to flicker. In 1946 Lilienthal and Fugitt[264] reported detecting the effects of carboxyhemoglobin concentrations of 5 to 9% when subjects had also been kept at a simulated pressure corresponding to an altitude of 1,524 or 1,829 m (5,000 or 6,000 ft). No such effect was observed at sea level, even at carboxyhemoglobin concentrations as high as 17%. The actual data were not reported; rather, the subjects' performance was characterized as "depressed" or "constant," which makes evaluation difficult. In another study reported the same year, Vollmer *et al.*[462] found no effect for carboxyhemoglobin concentrations of 22% with subjects at simulated altitudes of 4,724 m (15,500 ft). Fodor and Winneke in 1972,[141] Guest *et al.* in 1970,[178] O'Donnell *et al.* in 1971,[333] Ramsey in 1973,[365] and Winneke in 1974[491] all reported finding no effect on CFF at carboxyhemoglobin concentrations ranging from 10 to 12.7%.

Besides the auditory-vigilance and time-perception studies discussed above,[141,175,491] only one other study of the effects of small amounts of carbon monoxide on auditory function has been reported. In 1970 Guest *et al.*[178] used an auditory analogue of CFF, an auditory "flutter fusion" threshold. Their subjects were required to distinguish the point at which an intermittently presented sound was no longer heard as intermittent. They found that a carboxyhemoglobin concentration of 10% had no effect on this threshold.

The work by Guest *et al.*[178] is especially noteworthy in that it is one of the few studies with carbon monoxide that included within its design provision for gathering data on another agent with known effects, thereby providing for some internal validation of the sensitivity of its procedures.

Complex Intellectual Behavior

Dorcus and Weigand[126] used three tests of complex learned behavior in their 1929 study of the effects of automobile exhaust gas. No effects were found for carboxyhemoglobin concentrations up to 35%. In 1963 Schulte[396] studied firemen carrying out a series of complex tasks, none of which is described in detail. He was able to produce large and regular changes. For example, in one test subjects underlined all the plural nouns in certain prose passages. Between carboxyhemoglobin concentrations of 0 and 7% they averaged about 150 s to complete this task; between 10 and 15% they averaged about 210 s. Subjects averaged slightly less than 800 s to complete an arithmetic test at a concentration of 8%, but took about 1,000 s at 15%.

Both Stewart *et al.*[424] and Mikulka *et al.*[313] have suggested that Schulte's[396] measurements of carboxyhemoglobin concentrations were probably low since he reported a zero value under control conditions in a population consisting mainly of smokers. Both Guest *et al.*[178] and Stewart *et al.*[424] have questioned the reported concentrations of 20% after exposures to carbon monoxide at only 100 ppm. Stewart *et al.*[424] suggested the possibility that Schulte's[396] analytic techniques were unreliable.

O'Donnell *et al.*[333] also examined carbon monoxide's effects on ability to do a short series of mental arithmetic problems. Their four subjects took a mean of 89.8 s (SE = 10.56) after overnight exposure to 75 ppm and 98.6 s (SE = 11.52) after exposure to 150 ppm. The data from the control (0 ppm) measurements are not included here since they were always taken during the last of a series of three experimental sessions. The difference between the scores of the four subjects after the two different exposures (carboxyhemoglobin concentrations of 5.9 and 12.7%) is not reliable.

Bender *et al.*[41,42] investigated the effects of a moderate amount of carbon monoxide on several complex tasks. One of these was learning 10 meaningless syllables so that they could be recited without error. Exposure to 100 ppm carbon monoxide for about 2.5 hr (average carboxyhemoglobin concentration, 7%) produced a reliable decrease in accuracy. Repeating a series of digits in reverse order was also tested. It too showed a reliable decrease. Negative results were found for several other tasks involving calculation problems, analogies, shape selection, dot counting, and letter recognition.

In view of the recent surge of interest among psychologists in human perception, learning, and memory—all of which now go to make up cognitive psychology—as well as in human operant conditioning, it is surprising that more work has not been done on these topics with carbon monoxide. It would be illuminating, for instance, to see what effects carbon monoxide has upon such diverse behaviors as simple decision making[63] and the complex performance of aircraft pilots,[154,159] both of which have been shown to be sensitive to hypoxic hypoxia.

TABLE 5-10 Influence of the Presence of Carboxyhemoglobin on Maximal Aerobic Capacity ($\dot{V}O_{2max}$)[a]

No. of Subjects	Duration of Max Test,[b] min	% HbCO	% Decrease in $\dot{V}O_2$ max	Comments	Reference
2	3-5	31	32	Bolus plus supplementation	329
2	3-5	25	18	Bolus plus supplementation	329
8 (S, NS)	4-5	20.5	23	Bolus plus supplementation	461
16 (S, NS)	4-5	20.3	23	Bolus plus supplementation	460
10 (S, NS)	3-5	19.2	23	Bolus plus supplementation	135
5	2-3	15.4	15	Bolus	352
10 (S, NS)	3-5	7.1	9	Bolus plus supplementation	135
1	3-5	4.8	9	Bolus plus supplementation	135
7 (S)[c]	15	5.2	0	50 ppm for duration 25 C	366
7 (S)[c]	15	5.1	0	50 ppm for duration 35 C	366
10 (S)	21	4.5	3.0	50 ppm for duration 25 C	368
10 (S)	20	4.1	0	50 ppm for duration 35 C	129
10 (NS)	22	2.7	3.3	50 ppm for duration 25 C	367
10 (NS)	21	2.5	2.0	50 ppm for duration 35 C	79
9 (NS)[c]	20	2.3	3.4	50 ppm for duration 25 C	366
9 (NS)[c]	19	2.3	5.7	50 ppm for duration 35 C	366
4 (NS)	23	3.2	4.9	75 ppm for duration 25 C	200
4 (NS)	23	4.3	7.0	100 ppm for duration 25 C	200

[a] Significant decreases observed at 4.8 and greater carboxyhemoglobin [HbCO] levels. Subjects were all male. S = smokers; NS = nonsmokers.
[b] Duration of exercise time for NS to point of fatigue was consistently reduced in all tests when carbon monoxide was present in the ambient air.
[c] Middle-aged subjects; all others were younger adults.

EFFECTS OF CARBON MONOXIDE DURING EXERCISE

It has long been known that when carboxyhemoglobin exceeds 45%, the capacity to perform physical work is drastically reduced. The subjects studied by Chiodi *et al.*[79] were unable to carry out tasks that required low-to-moderate physical exertion when their carboxyhemoglobin was 40–45%. Attention has been directed recently to determining the influence of various carboxyhemoglobin concentrations on maximal aerobic power (maximal oxygen uptake*). Most of the relatively small number of experiments carried out to date have studied a limited population of healthy young males (see Table 5-10). The majority of these studies have induced the requisite carboxyhemoglobin concentration in their subjects by first exposing them to a relatively high concentration of carbon monoxide and then proceeding with the maximal aerobic capacity tests while administering supplementary carbon monoxide; alternatively, it was assumed that the carboxyhemoglobin concentration remained unchanged during the test. There have been only a few experiments in which the subjects breathed carbon monoxide at the low ambient concentrations encountered in urban areas. Some studies failed to separate the data from smoking and nonsmoking subjects.

Within the carboxyhemoglobin concentration range of 5–35% there is a linear relation between the decrease in aerobic capacity or power ($\dot{V}O_2$) and the carboxyhemoglobin concentration (Figure 5-24). This decrement in the maximal aerobic capacity can be predicted from the equation:

$$\% \text{ decrease } \dot{V}O_{2_{max}} = 0.91 \text{ [HbCO]} + 2.2.$$

Aerobic power was not significantly decreased in young nonsmokers until a carboxyhemoglobin concentration above 4% was attained. In these experiments the number of subjects studied was small. If greater numbers had been used, some significant differences would probably have been noted at lower carboxyhemoglobin concentrations. There were considerable differences between the responses of young smokers and nonsmokers, especially at lower carboxyhemoglobin concentrations. Even though smokers had higher carboxyhemoglobin concentrations than nonsmokers, they often did not show any decrease in aerobic capacity. The significance of these observations is not understood.

*Maximal aerobic power is the highest oxygen uptake an individual can attain during physical work breathing air at sea level (0 m).

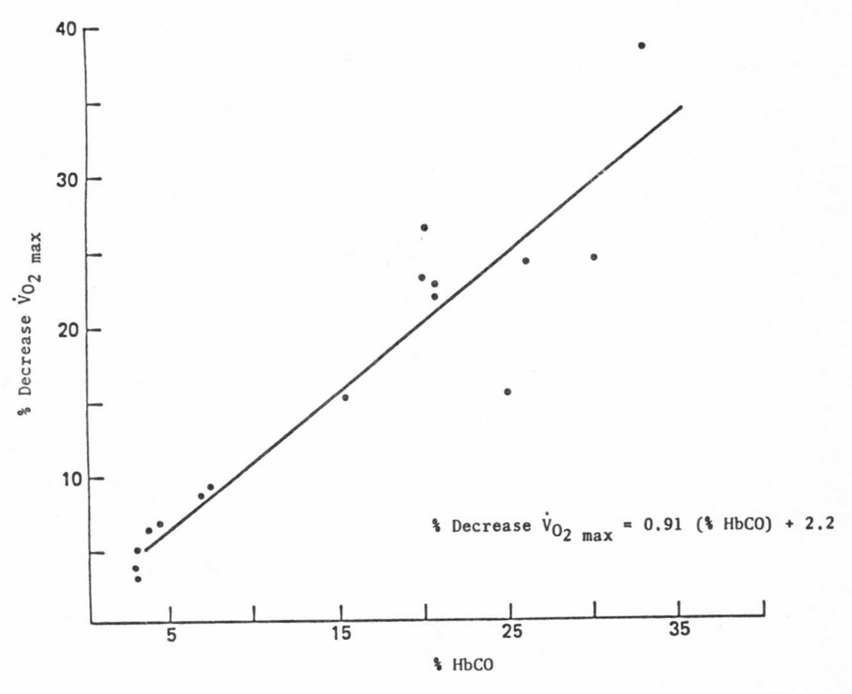

FIGURE 5-24 Relationship between percent carboxyhemoglobin and decrement in maximal aerobic power; the linear regression [% decrease $\dot{V}_{O_{2max}}$ = 0.91 (% HbCO) + 2.2] obtained only from 5 to 36% carboxyhemoglobin.

Too few middle-aged subjects were studied to permit drawing conclusions. In the studies reported[366,367] it was noted that these 41- to 56-yr-old men were not typical of the population. Because of the stringent testing, subjects were carefully screened to eliminate those with cardiovascular or pulmonary disabilities. Over half of the subjects were eliminated in the initial screening process. The data represent the responses of a highly special population. Therefore conclusions cannot be extended to the general middle-aged population, which has a risk factor related to the high incidence of cardiovascular disease. Furthermore, even though middle-aged smokers did not exhibit clinical signs of cardiopulmonary disability, they had a much lower aerobic capacity (in filtered air) than would have been predicted on the basis of age-related norms and so again are different from the general population. Changes in ambient temperature (25–35 C) appear to have only a minimal influence on aerobic capacity at the low carboxyhemoglobin concentration studied,[129,366] other than that anticipated on the basis of an increased thermal load.

Recent research[200] has indicated that 4.3% is a critical concentration at which carboxyhemoglobin reduces a maximal aerobic capacity. This was also accompanied by a reduction in total work time until maximal aerobic capacity was reached. Two procedures were used to raise the carboxyhemoglobin to appropriate concentrations: a buildup method in which carboxyhemoglobin was incrementally increased by administering ambient air containing carbon monoxide at 75 or 100 ppm CO, and a high initial carbon monoxide exposure followed by continued carbon monoxide inhalation to maintain the carboxyhemoglobin at the concentration reached in the buildup method, regardless of the magnitude of ventilation. The decrease in maximal aerobic capacity was found to occur at the same carboxyhemoglobin concentration and was therefore independent of the procedure followed. This observation indicated that even low ambient carbon monoxide concentrations (23.7 ppm) would result in a reduced maximal aerobic capacity if the individual had been previously exposed to sufficient carbon monoxide to raise his carboxyhemoglobin concentration to the critical value, 4.3%.[116] Clark and Coburn[84] have suggested that intracellular carbon monoxide effects may be responsible for the decrease in aerobic capacity.

The available data concerning the influence of various carboxyhemoglobin concentrations on the ability of young males to perform light to moderate work are summarized in Table 5-11. The only observable

TABLE 5-11 Influence of Carboxyhemoglobin on the Capacity to Perform Submaximal Work[a]

No. of Subjects	% HbCO	% Maximum Capacity[b]	Duration of Exercise, min	$\dot{V}O_2$ Uptake	Reference
3	25–33	40–68	60	No change	329
8 (S, NS)	23	50	8	No change	461
8 (S, NS)	23	75	8	No change	461
5	20	30	—	No change	135
5	20	70	—	No change	135
16 (S, NS)	20	45	7	No change	460
16 (S, NS)	20	70	7	No change	460
5	15	50	15	No change	352
8	15	50	13	No change	232
32 (S, NS)	14	75	5	No change	78
24 (S, NS)	3–6	35	240	No change	158

[a]S = smokers; NS = nonsmokers; HbCO = carboxyhemoglobin.

[b]These figures represent the work load at certain percentages of the subjects' maximal aerobic capacity, i.e., submaximal work.

effect is a slight increase in heart rate when the work is performed under conditions in which carboxyhemoglobin is increased. Because in all but one of the studies there was a minimal duration of effort (minutes) and relatively high percentages of maximal capacity, it will be necessary to separate these factors in order to interpret the influence of higher carboxyhemoglobin concentrations on such performance.

Studies conducted with both young (22–26 yr) and older (45–55 yr) subjects, smokers and nonsmokers, have indicated that at 35% of their maximal aerobic capacity prolonged periods of activity (3.5 hr within a 4-hr period) could be performed with minimal changes in their physiologic responses, even though carboxyhemoglobin concentrations were as high as 10.7 and 13.2% for nonsmokers and smokers, respectively[158] (see Table 5-12). However, heart rates were higher during this activity of walking while breathing polluted air (ambient carbon monoxide = 50, 75, or 100 ppm) than when walking while breathing carbon-monoxide-free air (0 ppm).

Previous investigations have indicated that a man could work at 35% of his maximal aerobic capacity for a period of 8 hr without evidence of physiologic stress, such as increasing heart rate. In this carbon monoxide study, the heart rate began to increase after 2 hr, indicating that the physiologic strain had started at an earlier time when carbon monoxide was present in the ambient air. Cardiac output remained constant but, since there was an increased heart rate, the heart beat

TABLE 5-12 Carboxyhemoglobin Concentrations Prior to and at Completion of 4 hr of Activity (35% Maximal Aerobic Capacity) at Different Concentrations of Ambient Carbon Monoxide[a]

	Nonsmokers[b]			Smokers[c]		
	% HbCO		Mean O_2	% HbCO		Mean O_2
Ambient CO, ppm	Pre	Post	Uptake, liters/min	Pre	Post	Uptake, liters/min[d]
0	0.63	0.32	1.29	4.64	1.80	0.85
50	0.67	4.88	1.31	6.23	6.88	0.72
75	0.85	10.27	1.26	5.48	10.68	0.86
100	0.78	12.56	1.21	4.41	13.18	0.94

[a] From Gliner *et al.*[158] Study consisted of two smokers and two nonsmokers. HbCO = carboxyhemoglobin.

[b] Minute ventilation during walking periods averaged 17.93 l.

[c] Minute ventilation during walking periods averaged 28.06 l.

[d] Smokers had significantly smaller maximal aerobic capacity values ($\dot{V}O_2$ max).

volume decreased. This alteration in volume was enhanced when work and carbon monoxide exposures were conducted in a warm environment (35 C).

In summary, maximal aerobic capacity is readily affected even at fairly low carboxyhemoglobin concentrations (5%), whereas submaximal efforts (30-75% of maximum) can be carried out with minimal changes in efficiency, even at relatively high carboxyhemoglobin concentrations (30%).

POPULATIONS ESPECIALLY SUSCEPTIBLE TO CARBON MONOXIDE EXPOSURE OWING TO REDUCED OXYGENATION AT ALTITUDES ABOVE SEA LEVEL

Precise data on the potential scope of the problems caused by carbon monoxide in residents and visitors at high altitiude are not available. There are approximately 2.2 million people living at altitudes above 1,524 m (5,000 ft) in the United States (see Table 5-13). Most live in nine states, with Colorado having the largest number. The majority of high-altitude residents (95%) live at 1,524-2,134 m (5,000-7,000 ft). The actual number of people exposed to carbon monoxide at these altitudes may be much larger because the tourist population in these states is high both in the summer and winter. Since proper tuning of automobiles for high-altitude traveling is uncommon, the influx of visitors with cars, accompanied by an increase in the emission of carbon monoxide and other contaminants, may be an important factor in increasing air pollution to an unacceptable point.

Ambient air standards set at sea level are not applicable to high-

TABLE 5-13 Estimated U.S. Population Living at High Altitudes (Nine States)[a]

Area	1,524-2,134 m (5,000-7,000 ft)	2,134-2,743 m (7,000-9,000 ft)	2,743-4,572 m (9,000-15,000 ft)
Urban	1,833,442	69,362	4,314
Rural[b]	321,100	16,770	1,150
Total	2,154,542	86,132	5,464

[a]Total number living above 1,524 m (5,000 ft) = 2,246,140 or approximately 6.7% of the total population of these nine states. Total U.S. population in 1970 was 203,235,000.

[b]Figures based on ratio of rural to urban population.

altitude sites. The Environmental Protection Agency's primary standards are expressed in milligrams per cubic meter of air. In Denver, at 1,524+ m (5,000+ ft), each cubic meter contains about 18% less air than at sea level. Therefore, permissible concentrations of carbon monoxide in Denver air would be 18% higher (a 10-mg/m^3 maximal permissible 8-hr average is equivalent to 8.7 ppm at sea level but 10.3 ppm at Denver's altitude). Table 5-14 illustrates the physiologically equivalent altitudes at various concentrations of carbon monoxide.

The effects of carbon monoxide and hypoxia induced by high altitude are similar. Carbon monoxide produces effects that aggravate the oxygen deficiency present at high altitudes. When high altitude and carbon monoxide exposures are combined, the effects are apparently additive. Each of these—a decrease in the partial pressure of oxygen in the air and increased carboxyhemoglobin—produce different physiologic responses. They have different effects on the partial pressure of oxygen in the blood, on the affinity of oxygen for hemoglobin, on the extent of oxyhemoglobin saturation (carbon monoxide hypoxemia shifts the oxyhemoglobin dissociation curve to the left, and a decrease in the alveolar oxygen partial pressure shifts it to the right), and on ventilation drive.

The actual influence of a combination of increased carboxyhemoglobin and decreased oxyhemoglobin has not been adequately documented by experimental data. The few available studies refer only to acute exposures to a decreased oxygen partial pressure and an increased carbon monoxide partial pressure. The best available information on the additive nature of this combination comes from psychophysiologic studies, but even they are inadequate. There are no data on the effects of carbon monoxide on residents at high altitudes or on their reactions when they

TABLE 5-14 Approximate Physiologically Equivalent Altitudes at Equilibrium with Ambient Carbon Monoxide Concentrations

Ambient Carbon Monoxide Concentration, ppm	Physiologically Equivalent Altitude at Actual Altitude of:					
	0 ft[a]	0 m[a]	5,000 ft	1,524 m	10,000 ft	3,048 m
0	0	0	5,000	1,524	10,000	3,048
25	6,000	1,829	8,300	2,530	13,000	3,962
50	10,000	3,048	12,000	3,658	15,000	4,572
100	12,300	3,749	15,300	4,663	18,000	5,486

[a]Sea level.

are suddenly returned to sea level and higher ambient carbon monoxide concentrations.

McFarland *et al.*,[300] in conjunction with their studies on the exposure of young males to high altitudes, showed that changes in visual threshold took place at carboxyhemoglobin concentrations as low as 5% or in a simulated altitude of approximately 2,438 m (8,000 ft). These observations were confirmed by Halperin *et al.*,[184] who further observed that recovery from detrimental effects on visual function lagged behind the carbon monoxide elimination. However, there were very few actual data given, and neither the variability among the four subjects nor the day-to-day variations were reported. Vollmer *et al.*[462] studied the effects of carbon monoxide at simulated altitudes of 3,048 and 4,572 m (10,000 and 15,000 ft) and did not observe any additive effects of carbon monoxide and altitude. They suggested that carbon monoxide's effects were masked by some compensatory mechanism. The data reported were not convincing.

Lilienthal and Fugitt[264] indicated that a combination of altitude 1,524 m (5,000 ft) and 5–9% carboxyhemoglobin caused a decrease in flicker fusion frequency, although either factor by itself had no effect. They also reported that a carboxyhemoglobin concentration of 8–10% reduced altitude tolerance by about 1,219 m (4,000 ft). Forbes *et al.*[143] found that during light activity at an altitude of 4,877 m (16,000 ft), carbon monoxide uptake was increased. This was probably the result of hyperventilation at high altitude caused by the respiratory stimulus of a decreased oxygen partial pressure. Pitts and Pace[353] stated that, if the subjects were at altitudes of 2,134 to 3,048 m (7,000 to 10,000 ft), every 1% increase in carboxyhemoglobin (up to 13%) was equivalent to a 108.2-m (355-ft) rise in altitude. Their observations were based on changes in the heart-rate response to work. These studies were contaminated in general by such factors as poor control and the presence of unidentified subjects who smoked.

Two groups of investigators have reported data comparing physiologic responses to high altitude and carbon monoxide where the hypoxemia due to altitude and the hypoxemia due to the presence of carboxyhemoglobin were approximately equivalent. In one study,[23] the mean carboxyhemoglobin concentration was approximately 12% (the method of carbon monoxide exposure resulted in a carboxyhemoglobin variation of from 5 to 20%) and the altitude was 3,454 m (11,333 ft). The second study[460] compared responses at an altitude of 4,000 m (13,125 ft) and a carboxyhemoglobin content of 20%. In both these studies, the carboxyhemoglobin concentration greatly exceeded that anticipated for typical ambient pollution. They both suggested, however, that the effects attributable to carbon monoxide and altitude were equivalent.

The precise measurement of the possible additive effects of carbon monoxide exposures and altitudes has not received much attention. What little information is available has been obtained by assuming simple additive effects,[202] but these have not been verified by direct experiments. In the construction of tunnels at 3,353 m (11,000 ft) it was recommended, on theoretical grounds, that the ambient carbon monoxide in the tunnel not exceed 25 ppm.[315] To elucidate this question research is needed following both physiologic and psychophysiologic approaches.

EFFECTS OF CHRONIC OR REPEATED CARBON MONOXIDE EXPOSURE

In the course of chronic or repeated carbon monoxide exposure over periods of several weeks or months, a variety of structural and functional changes develop. Some of these changes serve to offset the impairment brought about by carbon monoxide and thus can reasonably be taken to represent adaptation. Other changes, however, are of uncertain value to the organism, and a few are frankly disadvantageous. In this review, adaptive and nonadaptive changes are considered separately. The reader should recognize, however, that the separation is somewhat arbitrary and that further information may lead to reclassification.

Adaptation

The best evidence for adaptation is that animals chronically or repeatedly exposed to moderate concentrations of carbon monoxide can tolerate, without apparent harm, acute exposure to higher concentrations, which cause collapse or death in animals not exposed previously.[70,164,226,325,486] In addition, animals chronically exposed to carbon monoxide develop tolerance for acute exposure to simulated high altitudes, as well as the converse.[85] This suggests that common mechanisms are involved.

Whether there is symptomatic adaptation in man is much less clear; the evidence consists mainly of anecdotal reports of industrial workers exposed to unknown concentrations of carbon monoxide and of extensive studies conducted 40 yr ago by Killick, who herself was the subject.[227,228] She reported that in the course of several months of 6-hr exposures at 5- to 8-day intervals to carbon monoxide at 110–450 ppm, she became "acclimatized." This was manifested by diminished symptoms and a smaller pulse-rate response during exposure. She reported that after acclimatization, the equilibrium carboxyhemoglobin concentration reached during exposure to any given inspired carbon monoxide concentration was 30–50% lower than before. Because "acclimatized blood" equilibrated *in vitro* showed normal relative affinities for carbon monoxide and oxygen, and closed-circuit breathing experiments appeared to exclude the rapid metabolism of carbon monoxide, Killick suggested that after acclimatization carbon monoxide was actively transported from pulmonary capillary blood to alveolar air. It is unfortunate that this type of experiment has not been repeated with human subjects using modern analytical techniques. Both the results of animal experiments[53,257] and current evidence that pulmonary gas exchange occurs by passive diffusion suggest that there may have been a systematic technical error in Killick's data.

The mechanisms responsible for the development of symptomatic adaptation are not fully understood, but it is clear that hematologic changes are important. Chronic or repeated exposure to carbon monoxide causes an increase in both the hemoglobin concentration and the hematocrit (polycythemia) in a variety of experimental animals.[21,53,70,85,118,164,226,257,324,325,432,486] In most studies there is a rough correspondence between the severity of the carbon monoxide exposure and the extent of the polycythemic response. Man has a similar response, but neither the threshold nor the time course has been accurately quantitated. Industrial workers exposed to high but unmeasured amounts of carbon monoxide have been found to be significantly polycythemic.[212] Kjeldsen and Damgaard,[230] in a study of eight healthy volunteers exposed to 0.5% carbon monoxide intermittently for 8 days (mean carboxyhemoglobin concentration, 13%), found no change in hemoglobin or hematocrit values. This suggests that more severe or more prolonged exposure may be necessary to elicit this response in man. There is some evidence that cigarette smokers have higher hematocrits than nonsmokers, and, in a recent survey of blood donors, hemoglobin concentration was correlated with carboxyhemoglobin concentration in both smokers and nonsmokers.[418]

An increase in hemoglobin concentration increases the oxygen capacity of the blood and improves oxygen transport to some extent. The improvement may be limited owing to the increase in blood viscosity that accompanies the increased hematocrit.[323]

A second possible hematologic adaptation involves 2,3-diphosphoglycerate (2,3-DPG), a phosphorylated by-product of glycolysis that is found in the red blood cells of man and most other mammals. An increase in the concentration of 2,3-DPG shifts the oxygen-hemoglobin equilibrium in the direction of deoxygenation,[43,75] which lowers the effective oxygen affinity of hemoglobin (shifts the oxyhemoglobin dissociation curve to the right). This shift is of theoretical benefit during hypoxic stress because oxygen is "unloaded" into the tissues with a smaller drop in capillary oxygen partial pressure than would be possible with the normal dissociation curve.

Red cell 2,3-DPG concentrations are increased and the dissociation curve is shifted to the right in anemia and during residence at high altitudes.[223,253,341] Several investigators have looked for a similar effect from carbon monoxide exposuie, with inconclusive results. Dinman *et al.*[122] reported small increases in 2,3-DPG in humans after 3 hr at approximately 20% carboxyhemoglobin and in rats after exposure to variable higher concentrations of carbon monoxide. Conversely, Astrup[19] found a small decrease in red cell 2,3-DPG in human subjects maintained with 20% carboxyhemoglobin for 24 hr. Radford and Kresin[361] found that 2,3-DPG concentrations in rats were unaffected by 2- or 24-hr exposures to carbon monoxide at 500 and 1,000 ppm. Mulhausen *et al.*[321] studied blood from human subjects maintained for 8 days at an average carboxyhemoglobin concentration of 13% and found no change in the dissociation curve from that predicted from the immediate carbon monoxide effect. A shift of the dissociation curve does not appear to be an important adaptation to carbon monoxide exposure.

Except for Killick's[228] observation that her pulse rate at any given carboxyhemoglobin concentration was slower after acclimatization, there is little or no information about possible adaptation of the cardiovascular system. Data are not available about whether tissue capillarity increases with prolonged carbon monoxide exposure, as it does during high-altitude residence.[338,438] Muscle myoglobin concentration, especially in the heart, has been shown to increase at high altitude.[9,459] Although a similar increase might be expected owing to prolonged carbon monoxide exposure, no measurements have been made.

Montgomery and Rubin[317] in 1973 reported an example of adaptation. In a 1971 study[319] they demonstrated that exposing rats briefly (90 min) to carbon monoxide at 250–3,000 ppm resulted in a concentration-related slowing of the *in vivo* metabolism of certain drugs and a prolongation of their pharmacological effects. Hypoxia produced similar effects. When carbon monoxide exposure was prolonged, the effect on drug metabolism became less pronounced, and by 24 hr it had almost disappeared. Similar adaptation occurred during exposure to low inspired oxygen when hypocapnia was prevented. The mechanism of the adaptation is unknown.

Detrimental Effects of Chronic Exposure

The general category of chronic exposure effects comprises those that result from prolonged or repeated carbon monoxide exposure but which are not caused by acute exposure to carbon monoxide concentrations in the same range. Most of the effects discussed are irreversible or slowly reversible. The effects that do not appear to be of benefit to the organism—

and thus are not taken to represent adaptation—are in this section, but whether or not an effect is beneficial is questionable in some cases.

Prolonged exposure to carbon monoxide concentrations higher than encountered in ambient air pollution in the community environment has been shown to retard growth in experimental animals.[68,85,226] The mechanism of growth retardation has not been extensively studied, but reduced food intake may be involved.[236] The effects of carbon monoxide on fertility and on fetal development are reviewed elsewhere in this chapter.

A syndrome of chronic carbon monoxide intoxication has been described by several authors[38,153,177] but is far from being established. A wide variety of symptoms such as weakness, periodic loss of consciousness with twitching, insomnia, personality changes, loss of libido, and clinical and hematologic changes similar to those of pernicious anemia have all been attributed to chronic or repeated carbon monoxide exposure. Dogs exposed intermittently to carbon monoxide at 100 ppm during 11 weeks developed a broad-based gait and other subtle neurologic abnormalities,[257] but since there were no controls the results are questionable.

Cardiac enlargement, first reported in carbon-monoxide-exposed mice more than 40 yr ago,[69] has recently received renewed attention. Theodore *et al.*[439] reported an increase in heart weight in rats exposed at total pressure of 5 psi to 460 mg/m^3 of carbon monoxide for 71 days, followed by 575 mg/m^3 for 97 days. These unusual exposure conditions resulted in carboxyhemoglobin concentrations of 33–39% in dogs simultaneously exposed with the rats, but carboxyhemoglobin was not measured in the rats. Penney *et al.*[345,346] studied rats continuously exposed at sea-level pressure to carbon monoxide at 500 ppm (41% carboxyhemoglobin) and found that heart weight was significantly increased within a few days. After 2 weeks the heart weight was 35–40% greater than the value for control animals of the same body weight. Both the right and left ventricles were enlarged after 11 weeks of exposure. This contrasts with the predominance of right ventricular enlargement in rats exposed at high altitude. Heart weight was also increased in rats exposed to carbon monoxide for 30 days at 200 ppm (16% carboxyhemoglobin). In agreement with earlier reports,[133,324,427] animals that were exposed for 46 days to 100 ppm (9% carboxyhemoglobin) did not develop significant cardiac enlargement.

Histologic examination of myocardial tissue from animals exposed to 90–100 ppm of carbon monoxide for long periods has revealed edema, degeneration of muscle fibers, and fibrosis.[134,465] Kjeldsen *et al.*[231] in 1974 described a variety of ultrastructural changes in the hearts of rabbits exposed to carbon monoxide at 180 ppm (17% carboxyhemoglobin) for 2 weeks. The functional significance of these anatomic changes is not known.

During the past several years, Astrup *et al.*[21] have studied the influence of chronic carbon monoxide exposure on vessel walls. In cholesterol-fed rabbits, 8 weeks of exposure to carbon monoxide at 170 ppm (15–20% carboxyhemoglobin), followed by 2 weeks of exposure to carbon monoxide at 350 ppm (33% carboxyhemoglobin) increased the cholesterol content of the aorta and caused subendothelial edema. Similar effects were produced in normally fed monkeys exposed to carbon monoxide at 250 ppm (21% carboxyhemoglobin) for 2 weeks.[442] The significance of this observation in the pathogenesis of human vascular disease remains to be determined.

Significance for Human Health

Nearly all the available data on the effects of chronic CO exposure are derived from animal experiments. This is true both of adaptive changes and of effects that do not seem to be of benefit. Whereas many of these effects are of considerable interest, it is not justifiable at present to conclude that human beings are similarly affected. Existing data neither establish nor disprove a significant influence of chronic CO exposure on human health.

SUMMARY OF DOSE-RESPONSE CHARACTERISTICS IN MAN

This section summarizes present knowledge about the relationship of dose-response to adverse effects in man of acute carbon monoxide exposure.

Threshold for Adverse Carbon Monoxide Effects

Whether there is a threshold carboxyhemoglobin concentration for an adverse effect is still unknown. The question is of practical importance in setting carbon monoxide air standards. If there are adverse carbon monoxide effects at any carboxyhemoglobin concentration (no threshold), such effects could not be entirely prevented by legislation. The mechanism for adverse carbon monoxide effects is a fall in capillary oxygen partial pressure (pO_2) due to carbon monoxide binding to hemoglobin, and therefore a pertinent question is whether any fall in capillary pO_2, no matter how small, results in an adverse effect on tissues. It is known that many tissues, in order to keep intracellular pO_2 nearly constant, can adapt to acute falls in arterial pO_2 with resulting falls in capillary pO_2. The major adaptation mechanism in many tissues is probably recruitment of capillaries to give a decrease in oxygen diffusion distance between capillary blood and mitochondria. If such a mechanism occurs as carboxyhemoglobin increases, it is unlikely that adverse carbon monoxide effects occur at carboxyhemoglobin concentrations near zero and more probable that a threshold exists at a carboxyhemoglobin concentration where adaptation cannot compensate.

As indicated earlier in this chapter, the tissues most sensitive to the adverse effect of carbon monoxide appear to be heart, brain, and exercising skeletal muscle. Evidence has been obtained that carboxyhemoglobin concentrations in the 3–5% saturation range may adversely affect the ability to detect small unpredictable environmental changes (vigilance).[16,36,141,175] There is evidence that acute increases of carboxyhemoglobin to above 4–5% in patients with cardiovascular disease can exacerbate their symptoms[6,14,16] when the carboxyhemoglobin is as low as 5%.[199] Maximal oxygen consumption in exercising healthy young males has been shown to decrease when the carboxyhemoglobin is as low as 5%.[199] In the studies of the effect of carbon monoxide on vigilance and cardiovascular symptoms, there was no attempt either to determine the effect of lower carboxyhemoglobin concentrations or to look for a threshold. When aerobic metabolism of exercising skeletal muscle was studied, an apparent threshold was found. At a carboxyhemoglobin concentration below 5%, a measurable effect on oxygen uptake could not be demonstrated.

Dose–Response Relationships in Man

In recent years, the direction of research has been to look for adverse effects at low carboxyhemoglobin concentrations. Little effort has been made to investigate dose–response relationships at carboxyhemoglobin concentrations higher than those demonstrated to have an adverse effect. It is important to define dose–response relationships in order to determine whether an increase in carboxyhemoglobin will have the same adverse effect in a subject with a normal carboxyhemoglobin as it does in a subject with a higher baseline carboxyhemoglobin concentration, such as a smoker or a patient with an increased rate of endogenous carbon monoxide production. If the dose–response curve is concave upward, a given incremental increase in carboxyhemoglobin will have a greater adverse effect on a subject with higher baseline carboxyhemoglobin than on a subject with a normal one. Such a subject would have a greater risk of experiencing adverse effects from environmental carbon monoxide. The research mentioned above, that showed the effects of sudden small increases in carboxyhemoglobin on vigilance and on the myocardium, did not investigate the adverse effects over a wide range of carboxyhemoglobin concentrations. In the study of the effect of increasing carboxyhemoglobin on aerobic metabolism in exercising muscle (referred to above), it has been demonstrated that over a concentration range of 5 to 30% there was an almost linear relationship between carboxyhemoglobin and the fall in maximal oxygen uptake. This dose–response relationship cannot be extrapolated to carboxyhemoglobin effects on other tissues, since under conditions of exercise at maximal oxygen uptake it is unlikely that adaptation to carboxyhemoglobin could occur; the maximal recruitment of capillaries has probably already taken place because of exercise.

It is likely that dose–response relationships for the adverse effect of carbon monoxide on a given tissue are different in the presence of disease or in the fetus. Mechanisms of adaptation to carbon monoxide hypoxia may be altered or the tissue may be functioning under borderline hypoxia conditions and, therefore, be more susceptible to the effects of increased carboxyhemoglobin. Dose–response relationships may also be different when there is acute rather than chronic carbon monoxide exposure.

At the present time we know that when there are sudden increases in carboxyhemoglobin concentrations exceeding 20%, there are gross adverse effects on the functioning of several organ systems. Conversely, as discussed above, there is evidence of adverse effects on brain, myocardium, and skeletal muscle function at relatively low carboxyhemoglobin. But virtually no data are available in man for other dose–response relationships. Additional information about threshold carboxyhemoglobin concentrations and dose–response relationships for various tissues in normal and carbon-monoxide-susceptible populations would be valuable in understanding the biologic effects of carbon monoxide on man.

Ambient Air Standards for Carbon Monoxide

The current EPA standard for carbon monoxide is 9-ppm maximum for 8-hr average exposure, or 35-ppm maximum for 1-hr average exposure. Approximate calculated carbon monoxide uptakes for varying levels of activity after exposure to these concentrations are given below.

Exposure	Resting	Moderate Activity	Heavy Activity
9 ppm, 8 hr	1.3% sat	1.4% sat	1.4% sat
35 ppm, 1 hr	1.3%	2.2%	2.9%

These [HbCO] are calculated with the Coburn–Forster–Kane equation,[91] using appropriate values for carbon monoxide diffusing capacity, alveolar ventilation, alveolar pO_2, and endogenous carbon monoxide production for resting, moderate, or heavy activity. The current EPA standard is mainly justified on the basis of adverse carbon monoxide effects in patients with cardiac and peripheral vascular disease and effects of carbon monoxide on oxygenation of skeletal muscles in exercising normal human subjects. There appears to be an adequate safety factor between the lowest carboxyhemoglobin concentration that has been demonstrated to cause adverse effects and the maximal carboxyhemoglobin concentration that can occur at 9-ppm carbon monoxide for 8 hr or 35 ppm for 1 hr. However, the existing data base on the adverse effects of carbon monoxide exposure is not adequate to allow a precise setting of the carbon monoxide concentrations in ambient air owing to the uncertainties discussed throughout this report, and it is probable that, as more information becomes available, there can be justification for altering the present standards.

6

Effects on Bacteria and Plants

BACTERIA

The limited data available indicate that relatively high carbon monoxide concentrations can have an effect on certain airborne and soil bacteria. But there is no evidence that concentrations normally found in polluted atmospheres have an effect. Lighthart[262] studied carbon monoxide's effect on the survival of airborne bacteria. Vegetative cells of *Serratia marcescens* 8 UK, *Sarcina lutea*, and the spores of *Bacillus subtilis* var. *niger* were held in aerosols at 15 C and exposed to 85 ppm for up to 6 hr. The relative humidity (RH) was varied from 1 to 95%. At 88% RH and above, carbon monoxide appeared to protect the cells of aged aerosols of *Serratia marcescens* 8 UK from loss of viability, whereas below 75% RH the death rate was 4 to 20 times greater than in the control. Carbon monoxide's influence on the death rate of *Sarcina lutea* was much less than for *Serratia marcescens* 8 UK, but the death rate appeared to increase at a low RH and some protection was afforded at high RH. Below 85% RH the rate of loss of *Bacillus subtilis* var. *niger* viable spores was reduced.

Carbon monoxide prevents cytochrome oxidase in bacterial vegetative cells from coupling with oxygen in the electron transport system. It has been speculated that this interference with the energy system is responsible for the observed effects on the bacteria. It was suggested[145] that high carbon monoxide concentrations affect the cytochrome oxidase of an aquatic streptomycete by inhibiting some stages of the organism's growth.

The luminescence of some strains of aquatic bacteria has been reported to be affected by carbon monoxide.[413] Suspensions of bacteria spotted on luminescence sensor discs and exposed to atmospheres containing carbon monoxide gave a detectable response with concentrations as low as 3 ppm.

Lind and Wilson[265] reported that nitrogen fixation by *Azotobacter veinlandii* (a free-living nitrogen-fixing bacterium) was inhibited by carbon monoxide concentrations from 1,000 to 2,000 ppm and totally suppressed by concentrations from 5,000 to 6,000 ppm. Nitrogen fixation by *Nostoc muscorum* was inhibited by 1,000 ppm, and it approached complete inhibition with concentrations of 2,500 ppm.[58] Bergersen and Turner[47] reported that exposures to 86 ppm suppressed the nitrogen fixation of bacterial suspensions of *Rhizobium japonicum* obtained from soybean root nodules. Higher concentrations were required to inhibit nitrogen fixation by the intact nodules.

Soils exposed to high carbon monoxide concentrations develop the ability to convert it more rapidly.[206] This may be due to changes in the population of carbon-monoxide-assimilating organisms in the soil and indicates that carbon monoxide could have an effect on the ecology of some soil organisms. Both bacteria[235] and fungi[206] are known to be able to oxidize it.

PLANTS

Plants are relatively resistant to carbon monoxide. Much higher concentrations are required to cause injury or growth abnormalities than for pollutants such as sulfur dioxide, ozone, and hydrogen fluoride. For this reason data concerning its effects on plants are extremely limited, and most of the studies have been conducted with much higher concentrations than those to which plants are exposed in nature. Possible damage to vegetation could be caused in three ways: the production of leaf injury or growth abnormalities that reduce yield, growth, or quality; the suppression of nitrogen fixation in the soil or root nodule resulting in a deficiency of nitrogen for plant growth; and suppression in the rate of photosynthesis over a sufficiently long time period to cause a significant reduction in the plant's growth rate.

The available data indicate that carbon monoxide does not cause visible effects on plants at concentrations found in the ambient air, but at high concentrations it can produce various abnormalities. Knight *et al.*[238] measured a growth suppression of the etiolated epicotyl of sweet pea at 5,000 ppm. From this study they showed that ethylene, and not carbon monoxide, was the major toxic component of smoke. Zimmerman *et al.*[505,506] conducted extensive studies of carbon monoxide effects on plants. Their interest was not in carbon monoxide as an air pollutant but rather in finding a chemical that could induce root initiation and stimulate the growth of other plant parts. They exposed over 100 species to artificial atmospheres containing high carbon monoxide concentrations for periods of up to 23 days (most of the data presented are for 10,000 ppm). At 10,000 ppm they found a growth reduction in a number of the species. But the species varied widely both in their susceptibility to carbon monoxide and in their symptom expression. The most important responses observed at these high concentrations were leaf epinasty and hyponasty; leaf chlorosis; stimulation of the abscission of leaves, flower buds, and fruits; hypertrophied tissue on stems and roots; retardation of stem growth; reduction of leaf size; initiation of adventitious roots from young stem or leaf tissue; and modification of the natural response to gravity, causing the roots to grow upward out of the soil. Minina *et al.*[314] and the Heslop-Harrisons[193] found that cucumbers and hemp exposed during critical stages of development to carbon monoxide at 10,000 ppm induced a modification of sex expression, even to the point of total sex reversal in genetically male hemp plants. McMillan *et al.*[303] reported that a concentration of 20,000 ppm for 24 hr caused up to 100% leaf drop in certain geographic variants of *Acacia farneniana*. They did not find this response in the two other acacia species studied.

It has been shown that high carbon monoxide concentrations can inhibit nitrogen fixation of red clover plants and soybean root nodules,[47,266] but the available data do not indicate that ambient concentrations would have an adverse effect. Lind *et al.*[266] exposed red clover inoculated with *Rhizobium trifolii* to carbon monoxide for periods of up to 1 month. No effect was observed at 50 ppm, there was a 20% reduction in nitrate production at 100 ppm, and at 500 ppm nitrogen fixation was essentially halted. Carbon monoxide combines rapidly with the root nodule hemoglobin (leghemoglobin) at a rate 20 times faster than it combines with myoglobin[492] and apparently inhibits oxygen transport to the interior of the nodule.[46] Leghemoglobin facilitates oxygen diffusion into the interior of the nodule.[444] The oxygen concentration appears to be important for nitrogen fixation.[46]

Carbon monoxide also can limit bacterial nitrate production through inhibiting the enzymatic process of nitrogen fixation. Free-living nitrogen-fixing bacteria and bacteria isolated from soybean root nodules can be inhibited by high concentrations,[54,265] but Bergersen *et al.*[47] reported that higher concentrations were required to inhibit an intact soybean root nodule than to inhibit a bacterial suspension prepared from nodules. Inhibition of nitrogen fixation by carbon monoxide has in some cases been reported to be competitive[58,273] and in others to be noncompetitive.[204,265]

Bidwell and Fraser[49] reported a carbon-monoxide-induced growth suppression. In their studies on carbon monoxide uptake by detached leaves they measured a reversible reduction in photosynthesis in several plant species exposed to relatively low concentrations. For example, there was strong inhibition of the net carbon dioxide uptake rate (a measure of the growth rate) of grapefruit at a concentration of 1.6 ppm and complete inhibition at 7 ppm. These data indicate that photosynthesis might be inhibited at concentrations commonly measured in the atmosphere. If this is true, carbon monoxide could be one of the major pollutants measured in ambient air in or near large cities responsible for the suppression of plant growth.[203,307,434,441] These results reported by Bidwell and Fraser[49] have not been confirmed, and additional studies should be conducted.

7

Summary and Conclusions

PROPERTIES AND REACTIONS

Carbon monoxide is a colorless, odorless gas that is chemically stable at ordinary temperature and pressure. Its molecule is heteropolar, diatomic, and diamagnetic. It has a low electric dipole moment, short interatomic distance, and high heat of formation from atoms. These characteristics suggest that the molecule is a resonance hybrid of three valence structures. The poisonous nature of carbon monoxide is due to the strength of the coordination bond formed with the iron atom, which is stronger than that of oxygen in protoheme (a ferrous ion complex of protoporphyrin IX that constitutes part of the hemeprotein molecules that bind carbon monoxide).

Chemical reactions of carbon monoxide are critical to its formation or loss in the atmosphere. Its reactions with oxygen, water, nitrogen dioxide, ozone, atomic hydrogen, and organic free radicals are probably not important. A number of reactions with unstable intermediates, such as atomic oxygen and hydroxyl radicals, are important in atmospheric carbon monoxide chemistry. It is likely that reactions with atomic oxygen in the upper atmosphere play a role in oxidizing carbon monoxide. Hydroxyl radicals are generated in the atmosphere, particularly in air heavily polluted by automobile exhausts and in the natural atmosphere. The major reaction generating hydroxyl radicals is believed to be solar photolysis of aldehydes. Carbon monoxide is also formed during solar photolysis of aldehydes, but this is not an important source compared with carbon monoxide production from internal combustion of gasoline-air mixtures in the automobile. The generation of hydroxyl radicals is important, because they are known to react rapidly with carbon monoxide at ordinary temperatures. However, such trace contaminants as hydrocarbons, sulfur oxides, and nitrogen oxides will compete successfully for available hydroxyl radicals and thus reduce the significance of reaction with carbon monoxide. Theoretically, the rate of conversion of nitric oxide to nitrogen dioxide by hydroperoxyl radicals could be affected by ambient carbon monoxide concentrations, but carbon monoxide concentrations are affected to only a very small extent by nitric oxide conversion. Hydroxyl radical reaction with methane produces carbon monoxide, and this may be the most important natural source of carbon monoxide.

SOURCES AND SINKS

There is evidence that perhaps 10 times as much carbon monoxide is formed by natural processes as by processes related to the activities of man. The anthropogenic sources are due to incomplete combustion processes. The total anthropogenic carbon monoxide emission is much greater than the total anthropogenic emission of all other criteria pollutants. Anthropogenic carbon monoxide emission is estimated to have increased by about 28% in the period 1966–70. Global carbon monoxide emission from combustion sources was estimated at 359 million metric tons in 1970. Of the total anthropogenic carbon monoxide formed, the internal-combustion engine contributes by far the largest fraction. There was a great increase in carbon monoxide emission from 1940 to 1968, paralleling the increase in motor vehicles; since 1968, the emission has decreased, owing to installation of emission control devices. Other sources include industrial processes, agricultural burning, fuel combustion in stationary sources, and solid-waste disposal. It is estimated that 95% of the global anthropogenic carbon monoxide emission originates in the Northern Hemisphere.

Natural sources of carbon monoxide include volcanic activity, natural gases, photochemical degradation of such organic compounds as aldehydes, methane oxidation, and possibly solar photodissociation of carbon dioxide at high altitudes. Calculations of carbon monoxide formation by methane oxidation (based on estimations of hydroxyl radical concentration in the troposphere air) suggest this reaction contributes about one-fourth as much carbon monoxide as man-made sources. The oceans are also a significant source of carbon monoxide. Endogenous carbon monoxide production by man and animals is probably insignificant, compared with other carbon monoxide sources, in terms of total global carbon monoxide formation.

The average global background concentrations of 0.1 ppm carbon monoxide in air reflect a balance between formation and removal rate. Background carbon monoxide as low as 0.025 ppm is found in northern Pacific air and in the range of 0.04–0.08 ppm in nonurban air in California. Background carbon monoxide concentrations in unpolluted areas reflect the history of the air mass. The residence time for carbon monoxide in the atmosphere has been crudely estimated at approximately 0.2 yr.

If there were no removal of carbon monoxide, the average atmospheric concentration would increase at the rate of 0.06 to 0.5 ppm/yr. The sinks include oxidation by hydroxyl radicals (probably the major sink); biologic sinks, such as microorganisms in soil, vegetation, and metabolism in animals; and removal at the surfaces of such materials as charcoal and carbon.

ENVIRONMENTAL ANALYSIS AND MONITORING

There are many problems in the monitoring of any atmospheric pollutant that are related to variation in concentration and to analytic error. Carbon monoxide concentration can be shown to be variable on the time and space scale of the smallest atmospheric eddies and on all larger scales. Furthermore, for almost any pollutant, a monitoring device may well not record the same concentrations as are present at a receptor a few meters away. Carbon monoxide is taken up by the lungs of man very slowly; after an increase in inspired carbon monoxide concentration, it takes nearly 24 hr to reach an "equilibrium" blood carboxyhemoglobin content. The same is true after a decrease in inspired carbon monoxide concentration. Because carbon monoxide concentrations in a typical urban environment are variable in all three spatial dimensions, as well as in time, exposure of a typical highly mobile urban dweller to carbon monoxide will vary greatly in the course of a day's activity, and the relationship between the effects on human health and the concentrations measured at monitoring stations is complex. These complexities are partly responsible for the dearth of reliable epidemiologic data on the health effects of chronic carbon monoxide exposure. The situation is exacerbated by the general tendency of pollutants to vary together, owing to meteorologic factors.

Carbon monoxide concentrations in cities exceed background concentrations by at least an order of magnitude. Seasonal differences are small. Diurnal concentration follows diurnal traffic patterns and tends to peak during morning and evening rush hours. Concentrations decrease steeply with increasing altitude and are also affected by persistent air circulation patterns. Mathematical models have led to some success in computing concentrations at locations not covered by monitoring stations. The problem is made more difficult by the absence of a general theory of optimal placement of monitoring devices; anthropogenic emission is invariably concentrated, whereas natural emission is usually diffuse.

Carbon monoxide concentrations normally found in open air influence man and animals on a rather long-term basis. Since the turnover time or exchange rate, or biological half-life, is in hours in man, selection of brief measuring periods (averaging times) is not necessary.

There are a number of analytic methods for atmospheric carbon monoxide, but only the nondispersive infrared analyzers and the electrochemical analyzer (Ecolyzer) are in wide use. The method consisting of catalytic reduction of carbon monoxide to methane and gas chromatography with flame ionization detection is specific and sensitive, but it is inherently discontinuous, and most commercial units based on this method have not been highly dependable. Nondispersive infrared analyzers operate continuously, but they are significantly less sensitive and somewhat less specific. Specificity problems are reduced, but apparently not totally controlled, by gas pretreatment systems. Some of the data generated with the nondispersive infrared analyzers have been unreliable. Electrochemical methods (Ecolyzer) of monitoring carbon monoxide appear to be highly satisfactory in regard to reliability and precision.

Although blood carboxyhemoglobin is an almost unique biologic indicator of exposure to carbon monoxide air pollution, its accurate measurement at low concentrations is not easy and requires well-trained personnel. In addition, of course, if the plan is to sample the general public for carbon monoxide exposure, there is a practical limitation on the willingness of people to cooperate. For these reasons, use of blood samples for routine monitoring of carbon monoxide exposure of the general public is considered to be impractical.

The efficacy of carbon monoxide monitoring networks as indexes of the human health effects of this pollutant has not been demonstrated. It would be ideal to compare uptake, as determined by blood carboxyhemoglobin measurements, with measurements of carbon monoxide in air.

EFFECTS ON MAN AND ANIMALS

Carbon monoxide in the body may come from two sources: It may be endogenous, owing to the breakdown of hemoglobin and other heme-containing pigments; and it may be exogenous, owing to the inhalation of carbon monoxide. Endogenous carbon monoxide results in blood carboxyhemoglobin saturation of approximately 0.4% in a normal human. Uptake of exogenous carbon monoxide adds to this value. The process of uptake of exogenous carbon monoxide consists of inhalation, increase in alveolar carbon monoxide concentration, and diffusion through the pulmonary membrane and into the blood. Generally, the rates of diffusion and ventilation limit carbon monoxide uptake. Equations developed to describe carbon monoxide uptake in the lung include the following quantities: diffusing capacity of the lung, alveolar ventilation, oxygen tension in pulmonary capillary blood, and pulmonary capillary blood volume. Because the carbon monoxide diffusing capacity and pulmonary capillary blood volume vary with age and exercise, carbon monoxide uptake at a given inspired carbon monoxide concentration also varies with age and exercise. Body size also influences total body hemoglobin; with a larger pool of hemoglobin available for carbon monoxide binding, the rate of increase in carboxyhemoglobin will be smaller at a constant pulmonary uptake. Obviously, the health of the lung is important: With decreased diffusing capacity or alveolar ventilation, carbon monoxide uptake resulting from an increase in inspired-air carbon monoxide concentration or carbon monoxide excretion resulting from a decrease in inspired-air carbon monoxide concentration will be delayed. Barometric pressure is a factor because of the effect on inspired and alveolar pO_2, pulmonary capillary pO_2, and pCO at a given carbon monoxide concentration.

Several important questions are related to biologic effects of carbon monoxide. For example, will a smoker with a usual carboxyhemoglobin concentration resulting from inhalation of cigarette smoke have carboxyhemoglobin from ambient carbon monoxide simply additive to that present from smoking? The theory predicts that this would be the case. What is the influence of various parameters on the rate at which carboxyhemoglobin changes during the daily cycle of activity? Answers predicted from the theory are given in Chapter 5.

With regard to physiologic effects of carbon monoxide, the most important chemical characteristic of carbon monoxide is that, like oxygen, it is reversibly bound by hemoglobin and competes with oxygen for binding sites on the hemoglobin molecule. Because the affinity of hemoglobin for carbon monoxide is more than 200 times that for oxygen, carbon monoxide, even at very low partial pressures, can impair the transport of oxygen. That is, the presence of carboxyhemoglobin decreases the quantity of oxygen that can be carried to tissues and shifts the oxyhemoglobin dissociation curve to the left and changes the shape of this curve, so that capillary oxygen tensions in tissues are decreased; this is believed to hinder oxygen transport from blood into tissues. Although not proved, it is possible that carbon monoxide exerts deleterious effects by combination with the intracellular hemoproteins, myoglobin, cytochrome oxidase, and cytochrome P-450. The evidence is best for myoglobin: It has been demonstrated that the degree of carboxymyoglobin in saturation increases with increases in blood carboxyhemoglobin. Recent evidence that mitochondrial respiration is more sensitive to carbon monoxide during tissue hypoxia and that binding of carbon monoxide to myoglobin increases during tissue hypoxia at the same blood carboxyhemoglobin content suggests an intracellular mechanism of carbon monoxide toxicity under this condition.

Effects on the Fetus

It has recently become obvious that the fetus may be extremely susceptible to effects of carbon monoxide carried in maternal blood. The fetal carboxyhemoglobin content is chiefly a function of maternal carboxyhemoglobin and fetal endogenous carbon monoxide production, but in addition it is a function of placental carbon monoxide diffusing capacity and the factors that affect maternal carboxyhemoglobin content. Under steady-state conditions, fetal carboxyhemoglobin is about 10–15% greater than the corresponding maternal carboxyhemoglobin concentration. The rates of fetal uptake and elimination of carbon monoxide are relatively low compared with those of the mother. After a step change in inspired carbon monoxide concentration, the time for maternal carboxyhemoglobin to reach half its steady-state value is about 3 hr. In contrast, fetal carboxyhemoglobin requires about 7.5 hr to reach half its steady-state value, and final equilibrium is not approximated for 36–48 hr. Because of this lag in change in fetal carboxyhemoglobin and because the ratio of fetal to maternal carboxyhemoglobin is greater than unity under steady-state conditions, the mean fetal carboxyhemoglobin content is greater than that of the mother under a wide variety of circumstances.

Few studies have explored the effects of carbon monoxide on the growth and development of the embryo and fetus. Most of these used high carbon monoxide concentrations. The only study with a moderate carbon monoxide content showed decreased birth weight and increased neonatal mortality in rabbits.

As indicated above, carbon monoxide interferes with tissue oxygenation, both by decreasing the capacity of blood to transport oxygen and by shifting the blood oxyhemoglobin saturation curve to the left. Blood oxygen tension must therefore decrease to lower than normal before a given amount of oxygen will unload from hemoglobin. Thus, blood carboxyhemoglobin lowers tissue end-capillary or venous pO_2. This may result in tissue hypoxia if the pO_2 is below a critical point and tissue blood flow does not increase appropriately. Some theoretical effects of blood carboxyhemoglobin on tissue oxygenation and the effective changes in blood flow and arterial pO_2 values required to maintain normal oxygen delivery are reviewed in Chapter 5. These mechanisms may operate either individually or together to compromise oxygen delivery to developing cells. If present briefly at critical periods of embryonic or fetal development or if continued for long periods, these effects may interfere with normal development.

The hypoxic effects of carbon monoxide are similar to the hypoxic effects of high altitude; and the fetus, as well as the pregnant woman, at high altitude may be particularly sensitive to the effects of carbon monoxide.

The effects of carbon monoxide on fetal growth and development are of considerable interest and importance, but there is a dearth of experimental data available on them. Because of both the short-term effects of carbon monoxide on fetal oxygenation itself and the possible long-term sequelae (damage to the brain and central nervous system), fundamental research in this subject is urgent.

Maternal smoking is associated with increased blood carboxyhemoglobin in both mother and fetus. The decrease in mean birth weight of the infants of smoking mothers, compared with that of infants of normal nonsmokers, may result from relative hypoxia caused by carbon monoxide, but this is not established. Nicotine and other chemicals in tobacco may affect birth weight.

Much of the excessive exposure of the fetus and newborn infant to carbon monoxide results from smoking by the mother. There is considerable evidence that smoking during pregnancy results in increased incidences of abortion, such bleeding problems as placenta praevia and abruptio placentae, stillbirths, and neonatal deaths. It is thus apparent that there is no place for cigarette smoking during pregnancy (it must be noted that the results of smoking during pregnancy include not only the effects of carbon monoxide, but also the effects of nicotine and other constituents of tobacco smoke).

Cardiovascular Effects

It has become clear in recent years that the cardiovascular system, particularly the heart, is susceptible to adverse effects of carbon monoxide at low blood carboxyhemoglobin concentration.

Considerable evidence has been obtained with experimental animals that carboxyhemoglobin at 6-12% saturation results in a shift from aerobic to nonaerobic metabolism in the myocardium and that tissue oxygen tension may be compromised. In addition, there apparently are ultrastructural changes in the myocardium of experimental animals exposed to carbon monoxide that produce carboxyhemoglobin at 8-9% saturation for 4 hr. There is strong evidence that patients with coronary arterial disease are more susceptible to small increases in blood carboxyhemoglobin, in that their physiologic responses are different from those of normal subjects.

Arteriosclerotic heart disease is the leading cause of death and morbidity in the United States. Many asymptomatic people have extensive coronary atherosclerosis. Experimental and clinical studies have suggested that exposure to carbon monoxide is important in the development of atherosclerotic disease, later heart attacks, and the natural history of heart disease.

Persons with angina pectoris exposed to relatively low doses of carbon monoxide for short periods—doses that raised their hemoglobin content to about 2.5%—were found to be able to exercise for a shorter period before the onset of chest pain. Similarly, exposure to the air on the Los Angeles freeway resulted in an increase in carboxyhemoglobin concentration (mean, 5.08%) and a decrease in exercise tolerance among angina patients, compared with those in patients who breathed compressed air while driving on the freeway. Patients with "intermittent claudication"* who breathed carbon monoxide at 50 ppm for 2 hr developed leg pain sooner after the start of leg exercise and had greater duration of pain than when they had not been exposed to carbon monoxide.

Patients with angina pectoris who continue to smoke, and therefore might have higher carboxyhemoglobin contents than nonsmokers, have a poorer prognosis than patients with angina pectoris who do not smoke cigarettes.

After the cessation of cigarette smoking by people free of clinical coronary arterial disease, there is a reduction in their risk of heart attack.

*A complex of symptoms characterized by absence of pain or discomfort in a limb when at rest, the commencement of pain, tension, and weakness after walking is begun, intensification of the condition until walking becomes impossible, and the disappearance of the symptoms after a period of rest.

This reduction in risk occurs fairly rapidly after the cessation of smoking and suggests that cigarette smoking (and perhaps carbon monoxide) precipitates heart attack.

It is not known whether results obtained in studies of effects of carbon monoxide on patients with cardiovascular disease are pertinent to large numbers of people with these diseases in our population.

Exposure of selected animals, especially rabbits and primates, to carbon monoxide and a high-cholesterol diet has led to a higher incidence of atherosclerosis than a high-cholesterol diet alone. No data are available on the increase in atherosclerosis after carbon monoxide exposure in man. A recent report, however, noted atherosclerotic changes in the umbilical arteries of newborns of smoking mothers.

Epidemiologic studies in Baltimore and Los Angeles have failed to show any relationship between the number of heart attacks per day (including both myocardial infarction and sudden death) and the daily ambient carbon monoxide concentrations in the community. The case-fatality percentage was reported to be greater in areas within the Los Angeles basin with high ambient carbon monoxide content than areas with low content during the time when the ambient carbon monoxide in the basin was increased. Further analysis of these studies suggests, however, that other factors might have accounted for the difference between case-fatality percentages.

People who died suddenly from coronary arterial disease had higher postmortem carboxyhemoglobin concentration than other sudden-death victims and living controls. The differences were related primarily to the extent of cigarette smoking among the different groups. No difference in the pathologic findings between smoking and nonsmoking people who died suddenly from arteriosclerotic heart disease (ASHD) was noted in the Baltimore study.

Behavioral Effects

The behavioral effects of low concentrations of carbon monoxide are small and variable. The effects found most reliably in the laboratory are those on vigilance tasks, in which subjects are asked to report the occurrence of occasional signals over long periods. Four recent studies reported a higher incidence of missed signals at very low carboxyhemoglobin—between 2 and 5%—than under control conditions; but one study, in which the carboxyhemoglobin was 9%, did not. These studies all used relatively small numbers of young, healthy subjects. In addition, all but one relied on indirect measurement of carboxyhemoglobin, either alveolar breath samples or estimates made from knowledge of the duration of the exposure to carbon monoxide. Nevertheless, taken as a group, these studies argue that carbon monoxide does have an effect on human behavior at carboxyhemoglobin saturations even lower than those reached by chronic smokers.

The finding by Beard and Wertheim in 1967, that carboxyhemoglobin estimated at less than 5% is associated with deficits in a subject's ability to discriminate between short tones, still stands, but it has not yet been repeated by other workers in a fashion that would truly challenge it. But this finding now appears to be more relevant to questions of vigilance than to time perception, inasmuch as some investigators have minimized the role of boredom and fatigue and others have done the opposite—i.e., have excluded external influences and studied their subjects for relatively long periods. The former method has usually yielded negative results, and the latter, positive. Other tests of time perception have been almost uniformly negative, so time perception itself is probably not affected by very low carbon monoxide content.

Only the most tentative conclusion—that carbon monoxide has perhaps a slight deleterious effect on driving performance—can be drawn from the few preliminary attempts to study actual or simulated driving. The lack of clear-cut evidence is especially unfortunate in view of the importance of automobile drivers in any consideration of the behavioral effects of carbon monoxide.

Low concentrations of carbon monoxide may impair brightness discrimination; however, this finding dates to World War II and has not been adequately repeated. There are some hints that various verbal and arithmetic abilities and motor coordination are lessened by carbon monoxide, but the evidence is unsatisfactory.

Carbon monoxide may modify effects produced by other substances. People drive automobiles under the influence of sedatives, tranquilizers, alcohol, antihistamines, and other drugs. What would be innocuous amounts of such drugs if taken alone may become important determinants of behavior in the presence of low carboxyhemoglobin concentrations. Interactions with the other constituents of automobile exhaust may also occur.

Carbon Monoxide Effects during Exercise

Data on effects of increased carboxyhemoglobin during exercise indicate that aerobic capacity is compromised readily even at fairly low carboxyhemoglobin saturation, whereas submaximal efforts (30–75% of maximum) can be carried out with minor change in efficiency, even at relatively high carboxyhemoglobin saturation—i.e., 30%. The available data were obtained for the most part from studies on young male subjects.

Altitude and Carbon Monoxide Effects

Precise data on effects of carbon monoxide in high-altitude residents and visitors are not available. Some 2.2 million people live at altitudes above 5,000 ft in the United States. Ambient air standards set at sea level are not applicable for high-altitude sites. EPA primary standards are expressed in milligrams per cubic meter of air, and at high altitudes each cubic meter of space contains less air than at sea level; therefore, allowable carbon monoxide concentrations are higher. As noted in a previous section, carbon monoxide and oxygen are competitive, so adverse effects of carbon monoxide should occur at lower carbon monoxide concentrations if tissue pO_2 is less in subjects at high altitude. Carbon monoxide uptake during transient high carbon monoxide concentration will be more rapid, owing to the lower alveolar pO_2. There are some data that suggest that effects of carbon monoxide are additive to effects of hypoxia. The most important information on carbon monoxide exposures at altitude—the preciseness of potentially additive effects—has not received much attention, and what little information there is has not been verified by direct experiments.

Chronic Carbon Monoxide Exposure

Animals subjected to prolonged or repeated exposure to carbon monoxide at concentrations higher than those associated with community air pollution undergo adaptive changes, which enable them to tolerate acute carbon monoxide exposures that cause collapse or death in animals not previously exposed. Polycythemia develops in the course of chronic carbon monoxide exposure, and this probably contributes to the symptomatic adaptation. Other compensatory mechanisms are likely, but they have not been demonstrated. Limited evidence suggests that chronically exposed humans also develop symptomatic adaptation.

Prolonged carbon monoxide exposure also leads to detrimental effects, including growth retardation, cardiac enlargement, and an increased rate of development of atherosclerosis in some experimental animals. The extent to which similar changes occur in humans is unknown.

Dose-Response Relationships

Whether there is a threshold concentration of carboxyhemoglobin for an adverse effect is still unknown. The question is of practical importance in setting carbon monoxide air standards because, if there is no threshold and there are adverse carbon monoxide effects at any blood carboxyhemoglobin content, it is impossible to prevent all adverse effects by legislating air standards; however, more severe effects could be prevented.

As indicated in this report, the tissues most sensitive to adverse effects of carbon monoxide appear to be heart, brain, and exercising skeletal muscle. There is evidence that carboxyhemoglobin at 3–5% saturation has an adverse effect on ability to detect small unpredictable environmental changes (vigilance). There is evidence that acute increases in carboxyhemoglobin to above 5% saturation in patients with cardiovascular disease can result in exacerbation of symptoms. Carboxyhemoglobin as low as 5% saturation has been shown to decrease the maximal oxygen consumption during exercise in healthy young males. In the studies of effects on vigilance or cardiovascular symptoms, no attempt was made to determine effects of lower carboxyhemoglobin saturation and to look for a threshold. In the studies on aerobic metabolism of exercising skeletal muscle, an apparent threshold was found: Measurable effects on oxygen uptake could not be demonstrated with carboxyhemoglobin at less than 5% saturation, but it is possible that effects were present at lower carboxyhemoglobin saturation.

Recent research has been directed at adverse effects of low carboxyhemoglobin saturation, and little effort has been made to investigate dose–response relationships at carboxyhemoglobin saturations greater than that demonstrated to have an adverse effect. Defining dose–response relationships, however, is important in considering the question of whether increases in carboxyhemoglobin have the same adverse effects in a subject with a normal carboxyhemoglobin as in a subject with baseline increased carboxyhemoglobin (such as a smoker) or a patient with increased endogenous carbon monoxide production. The studies referred to above, showing effects of acute small increases in carboxyhemoglobin on vigilance or the myocardium, did not explore adverse effects over a wide range of carboxyhemoglobin saturation. It has been demonstrated in the study of effects of increasing carboxyhemoglobin on aerobic metabolism in exercising muscle that over about 5–30% saturation there was an almost linear relationship between carboxyhemoglobin and the fall in maximal oxygen uptake.

It is likely that dose–response relationships for adverse effects of carbon monoxide on a given tissue are different in the presence of disease. Mechanisms of adaptation to carbon monoxide hypoxia may be altered, or the tissue may operate under conditions of borderline hypoxia and therefore be more susceptible to effects of increased carboxyhemoglobin. Dose–response relationships in acute carbon monoxide exposure may be different from those in chronic exposure.

There are gross adverse effects on function in several organ systems with acute increases in carboxyhemoglobin to over 20% saturation. At the other end of the spectrum, there is evidence of adverse effects on brain, myocardium, and skeletal muscle function at relatively low carboxyhemoglobin. But almost no data on other dose-response relationships in man are available. The data base is inadequate for determination of air quality criteria standards for carbon monoxide with a reasonable degree of certainty.

A great deal is known about adverse effects of carbon monoxide on normal and abnormal man, and much of this knowledge supports the setting of rather stringent air carbon monoxide standards. The present EPA standards are designed to prevent carboxyhemoglobin over about 2.5% saturation; this gives an adequate safety factor, in that adverse effects of "acute" carbon monoxide exposures under experimental conditions are demonstrated at carboxyhemoglobin above 4-6% saturation.

This review made no attempt to survey the literature related to the possible role of carbon monoxide in the adverse effects of cigarette smoking. Carboxyhemoglobin in smokers is frequently greater than saturations that have been implicated as having adverse effects on normal or abnormal man. A major conclusion of this report is that it is imperative to determine the effects of carbon monoxide in smokers and to determine the possible role of carbon monoxide in the excess morbidity of cigarette smokers.

CARBON MONOXIDE EFFECTS ON BACTERIA AND PLANTS

Carbon monoxide reacts readily with cytochrome oxidase, and this reaction and the resulting effect on energy transport may be responsible for some of the observed effects of this pollutant on plants and bacteria. Carbon monoxide also reacts with leghemoglobin, affecting the oxygen transport system in the legume root nodule, and at least some plants appear to be able to metabolize carbon monoxide.

Carbon monoxide at 85 ppm can increase or decrease the survival rate of some airborne bacteria, depending on the relative humidity and the organism. Luminescence of some strains of marine bacteria can be inhibited by carbon monoxide when they are exposed to a carbon monoxide atmosphere on sensor disks.

Carbon monoxide at about 10,000 ppm can produce various leaf, stem, flower, and root abnormalities in higher plants. These abnormalities include retardation of growth, epinasty and chlorosis of leaves, abscission of leaves and other plant parts, initiation of adventitious roots from stem or leaf tissue, modification of the response to gravity, and modification of sex expression. Carbon monoxide at about 100 ppm can inhibit nitrogen fixation in root nodules, and carbon monoxide at about 1,000 ppm can inhibit nitrogen fixation of free nitrogen-fixing bacteria.

Inhibition of apparent photosynthesis (a measure of the growth rate) has been measured in excised leaves exposed to typical urban ambient concentrations of carbon monoxide. Carbon monoxide at 1-10 ppm inhibited the apparent photosynthetic rate of coleus, cabbage, grapefruit, and Phoenix palm. Inhibition of apparent photosynthesis, if it occurs in the field, is probably the only important effect of carbon monoxide on plants at ordinary concentrations.

CLOSING COMMENTS

We would be remiss if we did not reemphasize the important of the issues associated with exposure to carbon monoxide. Little research has been done on these problems, compared with that on many other important subjects. And the quality of this research has sometimes been less than excellent. Unfortunately, the public and legislatures often do not recognize that the roots of understanding of many of the problems of clinical relevance, such as the basic mechanism of the effects of carbon monoxide on the human organism, lie in fundamental research.

8

Recommendations

• *We recommend that studies be supported to determine the role of carbon monoxide in deleterious effects of cigarette smoking.*
The estimation of populations that are influenced by adverse effects of carbon monoxide is confounded by the lack of information about the cause of adverse effects of cigarette smoking and the possible role of carbon monoxide.

• *We recommend that effort be directed toward greater public awareness of the hazards of cigarette smoking during pregnancy.*
Although there is public awareness of the hazards of cigarette smoking related to lung cancer and other diseases, there has been little publicity about hazards to the fetus. There is growing evidence of serious deleterious effects of maternal cigarette smoking on the fetus.

• *We recommend expansion of the data base related to adverse effects of carbon monoxide on vigilance, on oxygenation of exercising skeletal muscle, and in atherosclerotic heart disease and peripheral vascular disease.* Carbon monoxide standards are now set on the basis of available data in these fields. Experiments in each of these fields have been performed on a relatively small number of human subjects. There is a need for replication of studies by other laboratories. Carbon monoxide exposure should be varied in duration and concentration. Dose–response relationships should be determined. Studies of adverse effects on vigilance and on oxygen uptake during exercise should be performed on susceptible populations. In studies on patients with atherosclerotic heart disease, it would be particularly useful to study: patients with extensive coronary arterial disease as determined by coronary angiography, the effects of positive exercise testing in people without clinical symptoms, and people with high risk of heart attack.

• *We recommend that studies be performed to determine whether heart, brain, and exercising skeletal muscle adapt to effects of small increases in blood carboxyhemoglobin (less than about 5–10% saturation).* The entire data base related to deleterious effects of carbon monoxide on the heart, brain, and exercising skeletal muscle was obtained from acute experiments. Yet exposure of the population to carbon monoxide can be chronic or intermittent. It is necessary to determine whether subjects can adapt and therefore become less susceptible to intermittent or chronic increases in carbon monoxide in ambient air. Susceptible populations should be studied.

• *We recommend rapid expansion of the data base relating physiologic and ambient carbon monoxide measurements, continuation of the recent approach of monitoring human exposure to carbon monoxide in urban communities with blood carboxyhemoglobin and alveolar carbon monoxide measurements, and acquisition by the* EPA *of a trained team capable of measuring blood carboxyhemoglobin.*
One of the uncertainties in evaluating effects of environmental carbon monoxide on health is the relationship of environmental exposure to carbon monoxide uptake. Existing methods of monitoring environmental carbon monoxide can be improved. The spacing patterns of individual measurement stations within a monitoring network need to be studied. Comparison of air monitoring data with blood carboxyhemoglobin measured either directly or as alveolar carbon monoxide concentration would allow study of the efficiency of air-monitoring systems.

• *We recommend an increase in the information on mechanisms of adverse carbon monoxide effects in man.*
It is not clear how very small decreases in hemoglobin oxygen-carrying ability and computed mean capillary pO_2 resulting from 3–5% carboxyhemoglobin can cause significant effects on tissue oxygenation. Studies of intracellular effects of carbon monoxide should be performed at 5–10% carboxyhemoglobin.

• *We recommend studies aimed at identifying susceptible populations.* Patients with respiratory insufficiency or anemia should particularly be studied.

• *We recommend research to determine the possible role of carbon monoxide in the increased incidence of abortion, stillbirth, and neonatal death associated with mothers who are heavy smokers and in the small-for-gestational-age infants of smoking mothers.*

• *We recommend research to determine fetal susceptibility to carbon monoxide.*
It is known that carbon monoxide can be concentrated in the fetal circulation and that blood oxygen tensions in the fetus are very low; these factors are expected to increase susceptibility to the adverse effects of carbon monoxide.

• *We recommend studies to determine whether increased carboxyhemoglobin is a factor in sudden deaths due to coronary arterial disease.* In previous studies, it was shown that people who died suddenly from coronary arterial disease had higher postmortem carboxyhemoglobin saturation than other sudden-death victims and living controls. The differences were related primarily to the amount of cigarette smoking in the ASHD subjects who died suddenly. No difference in the pathologic findings between smoking and nonsmoking people who died suddenly from ASHD was noted in the Baltimore study. Studies should be performed to assess whether these effects are in fact relevant to large numbers of urban people with coronary disease.

• *We recommend studies aimed at determining the relationship between carbon monoxide exposure in some industries and morbidity and mortality from heart disease.*
This information should give insight into the relationship between atherosclerotic heart disease and carbon monoxide. A better determination of industrial carbon monoxide exposure is needed. Measurement of either carboxyhemoglobin or expired-air carbon monoxide among employees in various "high-risk" industries should be completed.

• *We recommend further research to establish the extent of carbon-monoxide-induced decrements in vigilance.*
Much work remains before it will be possible to determine which aspects of performance are most sensitive to carbon monoxide. For example, decreasing the rate of signal presentation leads to poorer performance, as does increasing the rate of unwanted signals or increasing the length

of the experimental session. It would be illuminating to find out how carboxyhemoglobin saturation interacts with these variables. In addition, modern advances in psychophysics, particularly in signal-detection theory, promise to help to elucidate the effects of agents like carbon monoxide.

• *We recommend further studies of effects of increased carboxyhemoglobin saturation on driving performance.*
Automobile drivers probably constitute the most important target population when one considers the behavioral effects of carbon monoxide. The task of driving an automobile resembles a vigilance task in many ways. In the light of the findings on vigilance, studies on driving performance may uncover deleterious effects of low concentrations of carbon monoxide, provided that experimenters with sensitive methods study their subjects for a long enough period. Two main targets here are probably the long-distance truck driver, who performs a job that combines monotony with danger, and the taxi driver, who is continuously exposed to some of the highest urban carbon monoxide concentrations. An epidemiologic study of automobile accidents needs to be performed to determine the possible role of carbon monoxide.

• *We recommend studies aimed at elucidating possible adverse effects of carbon monoxide on sensory functions.*
Definitive work on the sensory effects of carbon monoxide has not yet been done, despite a history of more than 30 yr of sporadic effort. The important conflicts among early experimental findings were described by Lilienthal[263] 25 yr ago; they remain unresolved. Questions concerning complex intellectual behavior and motor coordination are also largely unanswered, after almost 50 yr of experimentation.

• *We recommend studies of interactions between carbon monoxide and other agents.*
Carbon monoxide may modify effects produced by other substances. People drive automobiles under the influence of sedatives, tranquilizers, alcohol, antihistamines, and other drugs. Amounts of such drugs that would be innocuous alone may become important determinants of behavior in the presence of low carboxyhemoglobin saturation. Interactions may also occur with the other constituents of automobile exhaust.

• *We recommend continuing research aimed at determining potentially additive effects of carbon monoxide and low oxygen tension.*
What little information is available has been obtained by assuming simple additive effects, but it has not been verified by direct experiments.

Studies involving both physiologic and psychophysiologic approaches are recommended in order to clarify this issue. This information will allow a more rational approach to setting carbon monoxide air standards at high altitude. Studies of effects of low carbon monoxide concentration on man, adapted and nonadapted to high altitude, are needed.

- *We recommend study of the effect of typical urban ambient carbon monoxide concentrations on several airborne bacteria.*
- *We recommend studies on intact plants to determine the degree of suppression of growth or apparent photosynthesis at different carbon monoxide concentrations in combination with other smog-induced reactions.*
- *We recommend more work related to natural sources and sinks of carbon monoxide.*

The natural sources and sinks of atmospheric carbon monoxide are not well understood. Carbon monoxide production occurs in soil, but little is known about it. Natural sources and sinks should be measured on a global basis.

- *We recommend that animal research be encouraged.*

It is at this level that the behavioral, as well as physiologic, mechanisms of action will most likely be ascertained. Emphasis on mechanisms of action is warranted, because such emphasis makes possible intelligent extrapolation to human behavior that may not lend itself to direct study. The recent growth of sophisticated animal psychophysics has made possible a concerted attack on questions involving sensory and perceptual effects of carbon monoxide in animals.

Appendix A:

Methods of Monitoring Carbon Monoxide

NONDISPERSIVE INFRARED SPECTROMETRY

In 1971 the Environmental Protection Agency (EPA) designated nondispersive infrared spectrometry (NDIR) as the reference method for continuous measurement of carbon monoxide.[454] This relatively reliable procedure that has been used for many years is based on the absorption of infrared radiation by carbon monoxide. First infrared radiation from an emitting source passes alternately through a reference and a sample cell and then through matched detector cells containing carbon monoxide. Since the carbon monoxide in the detector cells absorbs infrared radiation only at the characteristic frequencies of this compound, the detector becomes sensitive to those frequencies. The wall between the detector cells is a flexible diaphragm with an electrical position transducer attached to it. When there is a nonabsorbing gas in the reference cell, and carbon-monoxide-free air in the sample cell, the signals from the detectors are balanced electronically. The carbon monoxide introduced into the sample cell reduces the radiation reaching the sample detector, lowering the temperature and pressure in the detector cell and displacing the diaphragm. This electronically detected displacement is amplified

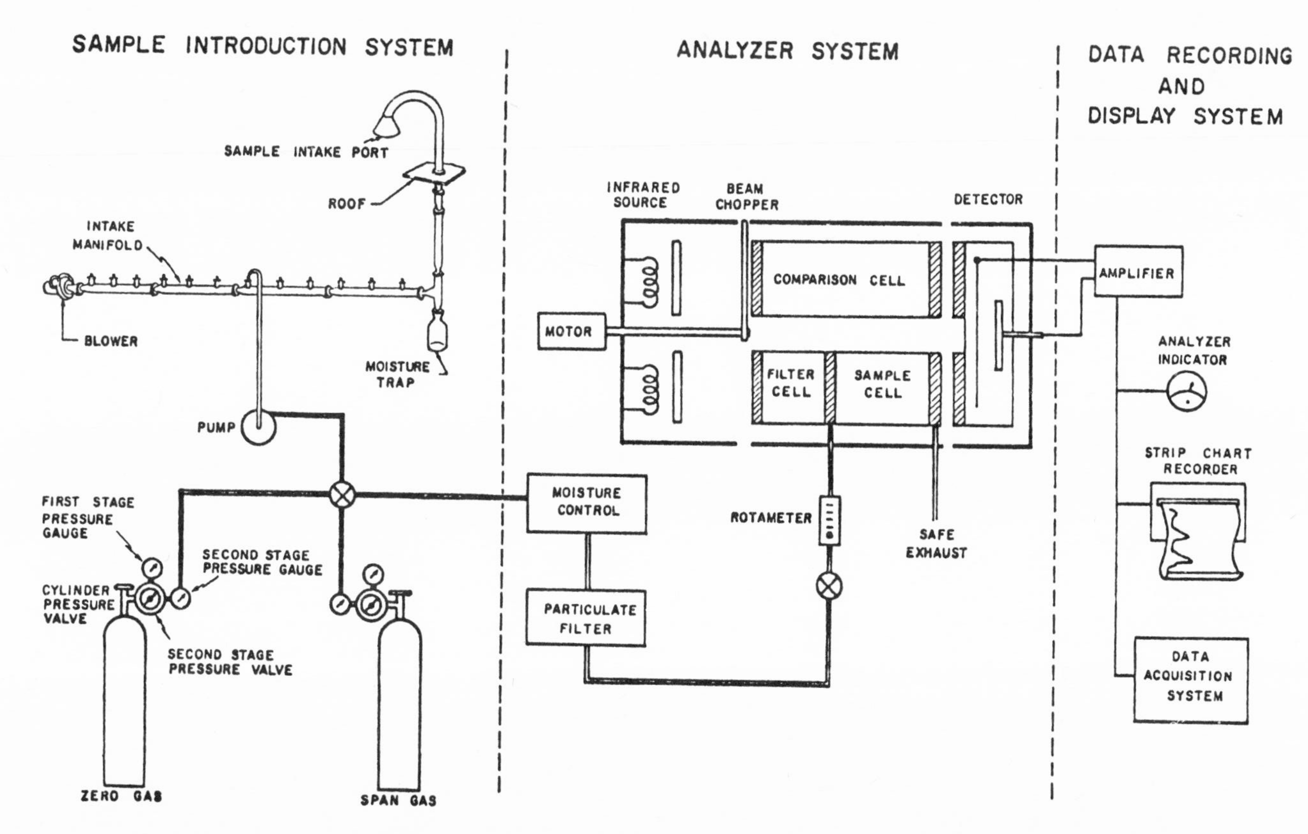

FIGURE A-1 Carbon-monoxide-monitoring system diagram. (From U.S. Environmental Protection Agency.[411])

to produce an output signal. Such a monitoring system for carbon monoxide is shown in Figure A-1.[411]

Because water is the principal interfering substance, the moisture control system is particularly important. Carbon monoxide data obtained using the nondispersive infrared technique are questionable without information about water removal. At 4.6 μm the absorption bands of water and carbon dioxide overlap the carbon monoxide band. A concentration of 2.5% by volume of water vapor can produce a response equivalent to 6.4 ppm carbon monoxide.[351] To reduce water vapor interference, water can be removed by drying agents or by cooling, or its effect can be reduced by optical filters. A combination of these is recommended. Selective ion exchange resins in a "heatless air drier" system can also be used. In a collaborative study of the NDIR method reported by McKee and Childers, a maximum reproducibility of ± 3.5 ppm in the 0 to 50 ppm range was found. The minimum detectable concentration was 0.3 ppm.[302] The instruments are large owing both to the long cells required for accuracy at low concentrations and to the air cooling and drying systems for water removal.

Since the data produced by the federal, state, and local monitoring agencies are used to make decisions which can be very costly, procedures for validating and maintaining data quality have been developed. Because of the requirements of the Clean Air Act, the EPA has issued specifications for instruments, methods, calibration, and data quality.[411] The performance specifications for automated carbon monoxide determinations are shown in Table A-1,[451] and the specifications for concentrations of interfering substances used to check the effects in automated analytical methods for carbon monoxide are summarized in Table A-2.[451]

DUAL ISOTOPE FLUORESCENCE

This instrumental method utilizes the slight difference in the infrared spectra of isotopes. The sample is alternately illuminated with the characteristic infrared wavelengths of carbon monoxide-16 (CO^{16}) and carbon monoxide-18 (CO^{18}). The carbon monoxide in the sample that has the normal isotope ratio, nearly 100% carbon monoxide-16, absorbs only the carbon monoxide-16 wavelengths. Therefore, there is a cyclic variation in the intensity of the transmitted light that is dependent on the carbon monoxide content of the sample.[268,291,292]

Full-scale ranges of 0 to 20 ppm and up to 0 to 200 with a claimed sensitivity of 0.2 ppm are available in this instrument. The response time (90%) is 25 s, but a 1-s response time is also available. An advantage of this technique is that it minimizes the effects of interfering substances.

TABLE A-1 Performance Specifications for Automated Analytical Methods for Carbon Monoxide[451]

Range	0–50 ppm
Noise	0.50 ppm
Lower detectable limit	1.0 ppm
Interference equivalent	
Each interfering substance	± 1.0 ppm
Total interfering substances	1.5 ppm
Zero drift	
12 hr	± 1.0 ppm
24 hr	± 1.0 ppm
Span drift, 24 hr	
20% of upper range limit	± 10.0%
80% of upper range limit	± 2.5%
Lag time	10 min
Rise time	5 min
Fall time	5 min
Precision	
20% of upper range limit	0.5 ppm
80% of upper range limit	0.5 ppm

Definitions:

Range: Nominal minimum and maximum concentrations that a method is capable of measuring.

Noise: The standard deviation about the mean of short duration deviations in output that are not caused by input concentration changes.

Lower Detectable Limit: The minimum pollutant concentration that produces a signal of twice the noise level.

Interference Equivalent: Positive or negative response caused by a substance other than the one being measured.

Zero Drift: The change in response to zero pollutant concentration during continuous unadjusted operation.

Span Drift: The percent change in response to an up-scale pollutant concentration during continuous unadjusted operation.

Lag Time: The time interval between a step change in input concentration and the first observable corresponding change in response.

Rise Time: The time interval between initial response and 95% of final response.

Fall time: The time interval between initial response to a step decrease in concentration and 95% of final response.

Precision: Variation about the mean of repeated measurements of the same pollutant concentration expressed as one standard deviation about the mean.

CATALYTIC COMBUSTION–THERMAL DETECTION

Determination of carbon monoxide by this method is based on measuring the temperature rise resulting from catalytic oxidation of the carbon monoxide in the sample air.

The sample air is first pumped into a furnace that brings it to a pre-

TABLE A-2 Test Concentrations of Interfering Substances for Automated Analytical Methods for Carbon Monoxide

	Concentration, ppm						
Measuring Principles	Ammonia	Nitric Oxide	Carbon Dioxide	Ethylene	Water Vapor	Methane	Ethane
Infrared photometric (other than reference method)	—	—	750	—	20,000	—	—
Gas chromatography (flame ionization detection)	—	—	—	—	20,000	—	0.5
Electrochemical	—	0.5	—	0.2	20,000	—	—
Catalytic combustion-thermal detection	0.1	—	750	0.2	20,000	5.0	0.5
Infrared fluorescence	—	—	750	—	20,000	—	0.5
Mercury replacement—ultraviolet photometric	—	—	—	0.2	—	—	0.5

set, regulated temperature and then over the catalyst bed in the furnace. A thermopile assembly measures the temperature difference between the air leaving the catalyst bed and the air entering the catalyst bed. The output of the thermopile, which is calibrated with known concentrations of carbon monoxide in air, is read on a strip-chart recorder as parts of carbon monoxide per million parts of air. The sensitivity is about 1 ppm. Most hydrocarbons are oxidized by the same catalyst, and will interfere unless removed.

ELECTROCHEMICAL

Carbon monoxide is measured by the means of the current produced in aqueous solution by its electro-oxidation at a catalytically active electrode. The concentration of carbon monoxide reaching the electrode is controlled by its rate of diffusion through a membrane. This is dependent on its concentration in the sampled atmosphere.[33,34] Proper selection of both the membrane and such cell characteristics as the nature of the electrodes and solutions make the technique selective for various pollutants. (A similar technique has been reported by Yamate *et al.*[498])

The generated current is linearly proportional to the carbon monoxide concentration from 0 to 100 ppm. A sensitivity of 1 ppm and a 10-s response time (90%) is claimed for a currently available commercial instrument.

Acetylene and ethylene are the chief interfering substances; 1 part acetylene records as 11 parts carbon monoxide and 1 part ethylene as 0.25 parts carbon monoxide. For hydrogen, ammonia, hydrogen sulfide, nitric oxide, nitrogen dioxide, sulfur dioxide, natural gas, and gasoline vapor, interference is less than 0.03 part carbon monoxide per 1 part interfering substance.

GAS CHROMATOGRAPHY–FLAME IONIZATION

Measured volumes of air are delivered 4 to 12 times/hr to a hydrogen flame ionization detector that measures the total hydrocarbon content (THC). A portion of the same air sample, injected into a hydrogen carrier gas stream, is passed through a column where it is stripped of water, carbon dioxide, and hydrocarbons other than methane. Methane is separated from carbon monoxide by a gas chromatographic column. The methane, which is eluted first, is unchanged after passing through a catalytic reduction tube into the flame ionization detector. The carbon monoxide eluted into the catalytic reduction tube is reduced to methane before passing through the flame ionization detector.[355] Between analyses the stripping column is flushed out. Nonmethane hydrocarbon concentrations are determined by subtracting the methane value from the total hydrocarbon value. There are two possible modes of operation. One of these is a complete chromatographic analysis showing the continuous output from the detector for each sample injection. In the other, the system is programmed for both automatic zero and span settings to display selected elution peaks as bar graphs. The peak height is then the measure of the concentration. The first operation is referred to as the chromatographic or "spectro" mode and the second as the barographic or "normal" mode.

Since measuring carbon monoxide entails only a small increase in cost, instrument complexity, and analysis time, these instruments are customarily used to measure three pollutants: methane, total hydrocarbons, and carbon monoxide.

The instrumental sensitivity for each of these three components (methane, total hydrocarbons, and carbon monoxide) is 0.02 ppm. The lowest full-scale range available is usually from 0 to 2 up to 0 to 5 ppm, although at least one manufacturer provides a 0 to 1 ppm range. Because of the complexity of these instruments, continuous maintenance by skilled technicians is required to minimize excessive downtime.

FRONTAL ANALYSIS

Air is passed over an adsorbent until equilibrium is established between the concentration of carbon monoxide in the air and the concentration of carbon monoxide on the adsorbent. The carbon monoxide is then eluted with hydrogen, reduced to methane on a nickel catalyst at 250 C, and determined by flame ionization as methane.

Concentrations of carbon monoxide as low as 0.1 ppm can be measured. This method does not give instantaneous concentrations but does give averages over a 6-min or longer sampling period.[130,131]

MERCURY REPLACEMENT

Mercury vapor formed by the reduction of mercuric oxide by carbon monoxide is detected photometrically by its adsorption of ultraviolet light at 253.7 nm. It is potentially a much more sensitive method than infrared absorption because the oscillator strength of mercury at 253.7 nm is 2,000 times greater than that of carbon monoxide at 4.6 μm.

Hydrogen and hydrocarbons also reduce mercuric oxide to mercury and there is some thermal decomposition of the oxide. Operation of the detector at constant temperature results in a regular background concentration of mercury from thermal decomposition. McCullough *et al.* recommended a temperature of 175 C to minimize hydrogen interference.[39,295] A commercial instrument employing these principles was made and used during the middle 1950's.[322] The technique has been recently used for measuring background carbon monoxide concentrations. Robbins *et al.*[374] have described an instrument in which the mercuric oxide chamber is operated at 210 C, and the amount of hydrogen interference was assessed by periodically introducing a tube of silver oxide into the intake air stream. At room temperature silver oxide quantitatively oxidizes carbon monoxide but not hydrogen. Thus the baseline hydrogen concentration can be determined. Additional minor improvements are discussed by Seiler and Junge,[398] who gave the detection limit for carbon monoxide as 3 ppb with a 5% standard deviation of the calibration at 0.2 ppm.

More recently Palanos[340] described a less sensitive model of this instrument intended for use in urban monitoring. It has a range of 0 to 20 ppm, a sensitivity of about 0.5 ppm, and a span and zero drift of less than 2% per day. As in other similar instruments, specificity is achieved by removal of the potentially interfering substances other than hydrogen (which is less than 10%).

All of these instruments assume a constant hydrogen concentration. In unpolluted atmosphere the hydrogen concentration is roughly 0.1 ppm. However, the automobile is not only a source of carbon monoxide but also of hydrogen. Therefore, if this technique is used in polluted areas, it will be necessary to measure the hydrogen concentration frequently.

Appendix B:

Measurement of Carbon Monoxide in Biologic Samples

Carbon monoxide is bound chemically to a number of heme proteins in the body, any of which could act as a measure of exposure. But because hemoglobin in the red blood cells binds carbon monoxide much more strongly in relation to oxygen than do any of the other proteins, and because blood is more easily obtained in a pure state than the other sources of heme proteins, blood is obviously the tissue of choice for sampling carbon monoxide exposure. Its affinity for hemoglobin is 220 times greater than that of oxygen,[213,380] about 40 times greater for myoglobin, and about the same for cytochrome P-450. For routine monitoring, however, blood samples can be taken only under special conditions. A more practical method for estimating carboxyhemoglobin is to measure alveolar gas.

The most meaningful expression for the evaluation of carbon monoxide uptake is the percent concentration of carboxyhemoglobin, [HbCO]. Alveolar carbon monoxide and the fraction of total hemoglobin unavailable for oxygen transport are both directly related to carboxyhemoglobin concentration. The carboxyhemoglobin concentration may be determined directly using spectrophotometric procedures without releasing the carbon monoxide bound with the hemoglobin. Also, it can be determined by measuring total hemoglobin separately and then measuring the amount of carbon monoxide present by liberating it as a gas. Venous blood can be taken by venipuncture or by pricking the earlobe or finger. The sample should be collected in a closed container containing an anticoagulant in the dry form, such as disodium ethylenediaminetetraacetic acid (EDTA) (1 mg/ml of blood) or dry sodium heparin USP (0.05 mg/ml of blood). Use of commercial anticoagulant Vacutainer tubes is satisfactory. Blood samples can be preserved on ice (about 4 C) for several days until analyzed. Methods that measure gaseous carbon monoxide and hemoglobin separately require complete mixing before aliquots are taken, which is not always easy with small blood samples.

CARBOXYHEMOGLOBIN MEASUREMENT

The amount of carbon monoxide in the blood can be determined by spectrophotometric procedures without liberating the bound carbon monoxide from the hemoglobin. These techniques are designed to give the percent of carboxyhemoglobin directly. Alternatively, the carbon monoxide content of the blood can be estimated by freeing all the carbon monoxide from the carboxyhemoglobin, extracting the gas, and assaying it by one of several techniques: classical volumetric methods, methods based on the reducing action of carbon monoxide, infrared absorption, or combinations of these. Many of the methods used are quite adequate when the carboxyhemoglobin percentage is greater than 20%. The difficulty has been to develop a method that is accurate for low concentrations of carboxyhemoglobin, particularly in the presence of other hemoglobin forms such as methemoglobin.

Spectrophotometric Methods

These methods have been popular because they are quick and simple, but they are frequently inaccurate at low concentrations. The spectrophotometric determination of carboxyhemoglobin is dependent on the difference between the carboxyhemoglobin absorption curve and the absorption curve for other forms of hemoglobin that are present at certain wavelengths of electromagnetic radiation. Spectrophotometry can be as simple and qualitative as observing the color of diluted blood. More precise and objective procedures are used to estimate the quantity of carboxyhemoglobin and the degree of saturation. A number of spectrophotometric procedures have been developed that vary in sophistication and accuracy. Klendshoj *et al.*[234] diluted blood 1:100 with dilute ammonia, added solid hydrosulfite, and measured the absorbance at both 555 and 480 nm. The addition of the hydrosulfite prevents presence of any other but the two pigments, carboxyhemoglobin and reduced hemoglobin. Both of these have the same absorbance at 555 nm but different absorbances at 480 nm. The ratio of absorbance, 555/480, decreases with increasing carboxyhemoglobin. It is evaluated using a standard curve prepared by analyzing known standards. This method, which is simple, rapid, and sufficiently accurate to determine a 2–5% change in carboxyhemoglobin concentration, has not proven satisfactory at low concentrations.

Small *et al.*,[409] using blood diluted about 1:70 in dilute ammonia, made absorbance measurements in the Soret region (410–435 nm) at four wavelengths with a 1-mm light path. A series of simultaneous equations were used to estimate the percent carboxyhemoglobin, the percent methemoglobin, and by difference the percent oxyhemoglobin. The accuracy of this is ±0.6% at low carboxyhemoglobin concentrations and ±2% methemoglobin at concentrations below 20%. This method has now been successfully adopted by other laboratories.

Probably the most convenient spectrophotometric procedure is automated differential spectrophotometry carried out with a carbon monoxide oximeter (manufactured by the Instrumentation Laboratory Co.) described by Malenfant *et al.*[287] In this method measurements of the three-component system containing reduced hemoglobin, oxyhemoglobin, and carboxyhemoglobin are made at three appropriate wavelengths: 548, 568, and 578 nm. The instrument carries out three simultaneous absorbance measurements at the three wavelengths on an automatically diluted, hemolyzed blood sample. The signals are then processed by an analog computer and displayed in digital form as total hemoglobin and the percent concentration of oxy- and carboxyhemoglobin. Although this instrument is commercially available and widely used, accurate measurements at low carboxyhemoglobin concentrations (less than 5%) can be carried out only after careful calibration.

Measurement of carboxyhemoglobin in blood using infrared spectroscopy of blood (rather than of carbon monoxide extracted from blood as described below) has been recently described.[290] This method appears to have a very high specificity for carbon monoxide bound to hemoglobin and a precision that is in the same range as that of the most precise methods in common use.

Volumetric Methods

A variety of methods are used to free bound carbon monoxide from hemoglobin before measuring the amount of released carboxyhemoglobin. Carbon monoxide liberation has been carried out by acidification with various acids (sulfuric, lactic, hydrochloric, acetic, phosphoric) either with or without the addition of oxidizing agents (potassium ferricyanide, potassium bi-iodate). The carbon monoxide released may be determined gasometrically[395,458] or by reaction with palladium chloride.[3,247]

There are three methods currently available that have sufficient accuracy and sensitivity to detect small changes in blood carbon monoxide content on the order of 0.02 vol% or less. Of these, the infrared analyzer[90,148] and the Hopcalite carbon monoxide oxidation meter[267] both require blood samples of 1 ml or more. In the infrared or Hopcalite analyzer, a major problem is the necessity for an accurate gas-phase dilution before analysis.

The most satisfactory current method is that of gas-solid chromatography on molecular sieve columns. Procedures employing thermal conductivity detectors[24,125] require a gas sample size of 1 ml or more, and the detectors must be operated at the highest sensitivity.

Porter and Volman[355] showed that hydrogen flame ionization detection can be used to measure carbon monoxide after in-line catalytic reduction to methane. This technique has been applied to the analysis of carbon monoxide both in blood and in respiratory gases[101] and is about 10^4 times more sensitive than other chromatographic techniques.

To calculate percent carboxyhemoglobin, total hemoglobin must be determined. This is done by reaction to form cyanmethemoglobin.[128] The most satisfactory procedure is to use Van Kampen and Zijlstra reagent,[457] taking precautions both to prevent the gradual loss of hydrogen cyanide from the acid reagent and to allow sufficient time for total conversion of the carboxyhemoglobin.[379]

The data derived from the analysis of carbon monoxide in the blood by various techniques are given in Table B-1.

Measurement of Alveolar Gas

The theory of measuring carboxyhemoglobin by measuring alveolar gas is based on the idea that under certain conditions the gas in the lungs will equilibrate with the blood. Measurements of the gas phase can then be applied to determine carboxyhemoglobin by use of the Haldane relationship:

$$[HbCO] = M\frac{pCO}{pO_2}[HbO_2]$$

To get pO_2, the oxygen partial pressure in the arterial blood can either be measured or assumed; pCO, the carbon monoxide partial pressure,

TABLE B-1 Comparison of Representative Techniques for the Analysis of Carbon Monoxide in Blood

Reference	Method	Sample Volume, ml	Resolution,[a] ml/dl	Sample Analysis Time, min	CV, %[b]
Gasometric					
409	Van Slyke	1.0	0.03	15	6
380	Syringe-capillary	0.5	0.02	30	2-4
Optical					
213	Infrared	2.0	0.006	30	1.8
395	Spectrophotometric	0.1	0.08	10	
287	CO-oximeter	0.4	0.10	3	
Chromatographic					
113	Thermal conductivity	1.0	0.005	20[c]	1.8
234	Flame ionization	0.1	0.002	20	1.8
247	Thermal conductivity	1.0	0.001	30	2.0
115	Thermal conductivity	0.25	0.006	3	1.7

[a] Smallest detectable difference between duplicate determinations.
[b] Coefficient variation. Calculated based on samples containing less than 2.0 ml CO per dl.
[c] Estimated from literature.

is the measured quantity, M is the Haldane constant (equal to 220 with very little individual variation[380] at physiologic pH); and oxyhemoglobin concentration is assumed to be approximately 1 − carboxyhemoglobin concentration.

The problem in using this method is to approximate equilibrium conditions when the gas compositon is actually changing at all times. Two maneuvers have been tried to achieve equilibrium conditions, rebreathing and breath holding. With rebreathing, the subject breathes in and out into a bag. A gas sample is taken after a specified number of breaths. Since oxygen is being continuously removed and carbon monoxide concentrations are a function of both the oxyhemoglobin present as well as the total volume of the lungs and bag, this is a complex system with equilibrium difficult to achieve. For this reason the rebreathing method has been replaced by the breath-holding method.[217]

Jones *et al.*[217] have shown that when a subject holds his breath, alveolar carbon monoxide concentration increases initially as carbon monoxide leaves the blood to equilibrate. As the alveolar oxygen falls, however, carbon monoxide is reabsorbed into the blood due to the fall of oxyhemoglobin and the Haldane relationship. Thus alveolar pCO will go through a maximum depending on the duration of breath holding. Jones *et al.*[217] concluded that 20 s was the optimum period of time for both practical and theoretical reasons. This technique is now standard.

The subject expires to residual volume, inspires maximally, holds his breath for 20 s, and then breathes out as far as possible. With the aid of a three-way valve, the first 500 ml of expirate is discarded, and the remaining gas is collected by turning the valve to an air-tight bag. The gas in the bag is then analyzed with standard gas analyzers. For field use the Ecolyzer has proved to be a rugged and reliable instrument.

Theoretically, based on the Coburn-Forster-Kane equation,[91] the slope of the graph relating the percent concentration of carboxyhemoglobin to alveolar pCO in ppm should be about 0.155 at sea level for carboxyhemoglobin percent concentration values equivalent to between 0 and 50 ppm, and progressively lower for higher concentrations. Various researchers have reported discrepancies in the results. Forbes *et al.*[143] found a ratio of 0.14, Sjostrand[408] found 0.16, and Ringold *et al.*[372] found 0.20. Smith (unpublished observations) showed that the slope was about 0.18 and did not decrease at higher carboxyhemoglobin percent concentrations, as would be expected, because equilibrium was reached less effectively.

This method cannot be used with persons that have chronic lung disease, in whom the alveolar gas composition tends to be extremely variable. Moreover, the subject's cooperation is essential. Despite these difficulties, skilled personnel can achieve reliable data.

References

1. Abramson, E., and T. Heyman. Dark adaptation and inhalation of carbon monoxide. Acta Physiol. Scand. 7:303–305, 1944.
2. Adams, J. D., H. H. Erickson, and H. L. Stone. Myocardial metabolism during exposure to carbon monoxide in the conscious dog. J. Appl. Physiol. 34:238–242, 1973.
3. Allen, T. H., and W. S. Root. An improved palladium chloride method for the determination of carbon monoxide in blood. J. Biol. Chem. 216:319–323, 1955.
4. Altshuller, A. P., and J. J. Bufalini. Photochemical aspects of air pollution: A review. Environ. Sci. Technol. 5:39–64, 1971.
5. Altshuller, A. P., and J. J. Bufalini. Photochemical aspects of air pollution: A review. Photochem. Photobiol. 4:97–146, 1965.
6. Anderson, E. W., R. J. Andelman, J. M. Strauch, N. J. Fortuin, and J. H. Knelson. Effect of low-level carbon monoxide exposure on onset and duration of angina pectoris. A study in ten patients with ischemic heart disease. Ann. Intern. Med. 79:46–50, 1973.
7. Anderson, J. G. The absolute concentration of OH ($X^2\pi$) in the earth's stratosphere. Geophys. Res. Lett. 3:165–168, 1976.
8. Andrews, J., and J. M. McGarry. A community study of smoking in pregnancy. J. Obstet. Gynaecol. Brit. Commonw. 79:1057–1073, 1972.
9. Anthony, A., E. Ackerman, and G. K. Strother. Effects of altitude acclimatization on rat myoglobin. Changes in myoglobin content of skeletal and cardiac muscle. Amer. J. Physiol. 196:512–516, 1959.
10. Aronow, W. S., J. Cassidy, J. S. Vangrow, H. March, J. C. Kern, J. R. Goldsmith, M. Khemka, J. Pagano, and M. Vawtek. Effect of cigarette smoking and breathing

carbon monoxide on cardiovascular hemodynamics in anginal patients. Circulation 50:340–347, 1974.
11. Aronow, W. S., J. Dendinger, and S. N. Rokaw. Heart rate and carbon monoxide level after smoking, high-, low-, and non-nicotine cigarettes. Ann. Intern. Med. 74:697–702, 1971.
12. Aronow, W. S., J. R. Goldsmith, J. C. Kern, J. Cassidy, W. H. Nelson, L. L. Johnson, and W. Adams. Effect of smoking cigarettes on cardiovascular hemodynamics. Arch. Environ. Health 28:330–332, 1974.
13. Aronow, W. S., C. N. Harris, M. W. Isbell, S. N. Rokaw, and B. Imparato. Effect of freeway travel on angina pectoris. Ann. Intern. Med. 77:669–676, 1972.
14. Aronow, W. S., and M. W. Isbell. Carbon monoxide effect on exercise-induced angina pectoris. Ann. Intern. Med. 79:392–395, 1973.
15. Aronow, W. S., and S. N. Rokaw. Carboxyhemoglobin caused by smoking non-nicotine cigarettes. Effects in angina pectoris. Circulation 44:782–788, 1971.
16. Aronow, W. S., E. A. Stemmer, and M. W. Isbell. Effect of carbon monoxide exposure on intermittent claudication. Circulation 49:415–417, 1974.
17. Asmussen, I., and K. Kjeldsen. Intimal ultrastructure of human umbilical arteries. Observations on arteries from newborn children of smoking and nonsmoking mothers. Circ. Res. 36:579–589, 1975.
18. Åstrand, I., P. Övrum, and A. Carlsson. Exposure to methylene chloride. I. Its concentration in alveolar air and blood during rest and exercise and its metabolism. Scand. J. Work Environ. Health 1:78–94, 1975.
19. Astrup, P. Intraerythrocytic 2,3-diphosphoglycerate and carbon monoxide exposure. Ann. N.Y. Acad. Sci. 174:252–254, 1970.
20. Astrup, P. Some physiological and pathological effects of moderate carbon monoxide exposure. Brit. Med. J. 4:447–452, 1972.
21. Astrup, P., K. Kjeldsen, and J. Wanstrup. Enhancing influence of carbon monoxide on the development of atheromatosis in cholesterol-fed rabbits. J. Atheroscler. Res. 7:343–354, 1967.
22. Astrup, P., H. M. Olsen, D. Trolle, and K. Kjeldsen. Effect of moderate carbon-monoxide exposure on fetal development. Lancet 2:1220–1222, 1972.
23. Astrup, P., and H. G. Pauli, Eds. A comparison of prolonged exposure to carbon monoxide and hypoxia in man. Scand. J. Clin. Lab. Invest. 24(Suppl. 103):1–71, 1968.
24. Ayres, S. M., A. Criscitiello, and S. Giannelli, Jr. Determination of blood carbon monoxide content by gas chromatography. J. Appl. Physiol. 21:1368–1370, 1966.
25. Ayres, S. M., S. Giannelli, Jr., and H. Mueller. Effects of low concentrations of carbon monoxide. Part IV. Myocardial and systemic responses to carboxyhemoglobin. Ann. N.Y. Acad. Sci. 174:268–293, 1970.
26. Baardsen, E. L., and R. W. Terhune. Detection of OH in the atmosphere using a dye laser. Appl. Phys. Lett. 21:209–211, 1972.
27. Baker, F. D., and C. F. Tumasonis. Carbon monoxide and avian embryogenesis. Arch. Environ. Health 24:53–61, 1972.
28. Baker, F. D., C. F. Tumasonis, and J. Barron. The effect of carbon monoxide inhalation on the mixed-function oxidase activity in the chick embryo and the adult mouse. Bull. Environ. Contam. Toxicol. 9:329–336, 1973.
29. Barth, C. A., A. I. Stewart, C. W. Hord, and A. L. Lane. Mariner 9 ultraviolet spectrometer experiment: Mars airglow spectroscopy and variations in Lyman alpha. Icarus 17:457–468, 1972.
30. Bartlett, D., Jr. Pathophysiology of exposure to low concentrations of carbon monoxide. Arch. Environ. Health 16:719–727, 1968.

31. Bates, D. R., and A. E. Witherspoon. The photo-chemistry of some minor constituents of the earth's atmosphere (CO_2, CO, CH_4, N_2O). Monthly Not. Roy. Astron. Soc. (Lond.) 112:101–124, 1952.
32. Baulch, D. L., D. D. Drysdale, and A. C. Lloyd. Critical Evaluation of Rate Data for Homogeneous, Gas Phase Reactions of Interest in High-Temperature Systems, No. 1. Leeds University, England, May 1968.
33. Bay, H. W., K. F. Blurton, H. C. Lieb, and H. G. Oswin. Electrochemical measurement of carbon monoxide. Amer. Lab. 4(7):57–61, 1972.
34. Bay, H. W., K. F. Blurton, J. M. Sedlak, and A. M. Valentine. Electrochemical technique for the measurement of carbon monoxide. Anal. Chem. 46:1837–1839, 1974.
35. Beard, R. R., and N. Grandstaff. Carbon monoxide and human functions, pp. 1–26. In B. Weiss and V. G. Laties, Eds. Behavioral Toxicology. New York: Plenum Press, 1975.
36. Beard, R. R., and N. Grandstaff. Carbon monoxide exposure and cerebral function. Ann. N.Y. Acad. Sci. 174:385–395, 1970.
37. Beard, R. R., and G. A. Wertheim. Behavioral impairment associated with small doses of carbon monoxide. Amer. J. Public Health 57:2012–2022, 1967.
38. Beck, H. G. The clinical manifestations of chronic carbon monoxide poisoning. Ann. Clin. Med. 5:1088–1096, 1927.
39. Beckman, A. O., J. D. McCullough, and R. A. Crane. Microdetermination of carbon monoxide in air. A portable instrument. Anal. Chem. 20:674–677, 1948.
40. Behrman, R. E., D. E. Fisher, and J. Paton. Air pollution in nurseries: Correlation with a decrease in oxygen-carrying capacity of hemoglobin. J. Pediatr. 78:1050–1054, 1971.
41. Bender, W., M. Göthert, and G. Malorny. Effect of low carbon monoxide concentration on psychological functions. Staub-Reinhalt. Luft (Engl. Ed.) 32(4):54–60, 1972.
42. Bender, W., M. Göthert, G. Malorny, and P. Sebbesse. Wirkungsbild neidriger Kohlenoxid-Konzentration beim Menschen. Arch. Toxikol. 27:142–158, 1971.
43. Benesch, R., and R. E. Benesch. The effect of organic phosphates from the human erythrocyte on the allosteric properties of hemoglobin. Biochem. Biophys. Res. Comm. 26:162–167, 1967.
44. Benesch, W., M. Migeotte, and L. Neven. Investigations of atmospheric CO at the Jungfraujoch. J. Opt. Soc. Amer. 43:1119–1123, 1953.
45. Benson, F. B., J. J. Henderson, and D. E. Caldwell. Indoor-Outdoor Air Pollution Relationships: A Literature Review. Publ. AP-112. Research Triangle Park, N.C.: U.S. Environmental Protection Agency, 1972. 73 pp.
46. Bergersen, F. J. The effects of partial pressure of oxygen upon respiration and nitrogen fixation by soybean root nodules. J. Gen. Microbiol. 29:113–125, 1962.
47. Bergersen, F. J., and G. L. Turner. Comparative studies of nitrogen fixation by soybean root nodules, bacteroid suspensions and cell-free extracts. J. Gen. Microbiol. 53:205–220, 1968.
48. Bernard, C. Leçons sur les Effets des Substances Toxiques et Médicamenteuses. Paris: J.-B. Baillière et Fils, 1857. 488 pp.
49. Bidwell, R. G. S., and D. E. Fraser. Carbon monoxide uptake and metabolism by leaves. Can. J. Bot. 50:1435–1439, 1972.
50. Birnstingl, M., L. Hawkins, and T. McEwen. Experimental atherosclerosis during chronic exposure to carbon monoxide. Eur. Surg. Res. 2:92–93, 1970. (abstract)
51. Blackburn, H., P. Canner, W. Krol, S. Tominaga, and J. Stamler. The natural history of myocardial infarction in the coronary drug project. Prognostic indicators

following infarction, pp. 54–64. In G. Tibblin, A. Keys and L. Werkö, Eds. Preventive Cardiology. Proceedings of an International Symposium held at Billingehus, Skövde, Sweden, Aug. 21, 1971. New York: Halsted Press, 1972.

52. Brief, R. S., R. S. Ajemian, and R. G. Confer. Iron pentacarbonyl: Its toxicity, detection, and potential for formation. Amer. Ind. Hyg. Assoc. J. 28:21–30, 1967.
53. Brieger, H. Carbon monoxide polycythemia. J. Ind. Hyg. Toxicol. 26:321-327, 1944.
54. Brody, J. S., and R. F. Coburn. Effects of elevated carboxyhemoglobin on gas exchange in the lung. Ann. N.Y. Acad. Sci. 174:255–260, 1970.
55. Brown, F. B., and R. H. Crist. Further studies on the oxidation of nitric oxide; the rate of reaction between carbon monoxide and nitrogen dioxide. J. Chem. Phys. 9:840–846, 1941.
56. Buncher, C. R. Cigarette smoking and duration of pregnancy. Amer. J. Obstet. Gynecol. 103:942–946, 1969.
57. Burns, B., and G. H. Gurtner. A specific carrier for oxygen and carbon monoxide in the lung and placenta. Drug Metab. Depos. 1:374–379, 1973.
58. Burris, R. H., and P. W. Wilson. Characteristics of the nitrogen-fixing enzyme system in *Nostoc muscorum.* Bot. Gazette 108:254–262, 1946.
59. Butler, N. R., and E. D. Alberman, Eds. Perinatal Problems. The Second Report of the 1958 British Perinatal Mortality Survey, under the Auspices of the National Birthday Trust Fund. Edinburgh: E. & S. Livingston, Ltd., 1969.
60. Butler, N. R., and D. G. Bonham. Perinatal Mortality. The First Report of the 1958 British Perinatal Mortality Survey, under the Auspices of the National Birthday Trust Fund. Edinburgh: E. & S. Livingston, Ltd., 1963. 308 pp.
61. Butler, N. R., and H. Goldstein. Smoking in pregnancy and subsequent child development. Brit. Med. J. 4:573–575, 1973.
62. Butler, N. R., H. Goldstein, and E. M. Ross. Cigarette smoking in pregnancy: Its influence on birth weight and perinatal mortality. Brit. Med. J. 2:127–130, 1972.
63. Cahoon, R. L. Simple decision making at high altitude. Ergonomics 15:157–164, 1972.
64. Calvert, J. G., K. L. Demerjian, and J. A. Kerr. Computer simulation of the chemistry of a simple analogue to the sunlight-irradiated autoexhaust polluted atmosphere. Environ. Lett. 4:123–140, 1973.
65. Calvert, J. G., K. L. Demerjian, and J. A. Kerr. The effect of carbon monoxide on the chemistry of photochemical smog systems. Environ. Lett. 4:281–295, 1973.
66. Calvert, J. G., J. A. Kerr, K. L. Demerjian, and R. D. McQuigg. Photolysis of formaldehyde as a hydrogen atom source in the lower atmosphere. Science 175:751–752, 1972.
67. Calvert, J. G., and J. N. Pitts, Jr. [The oxides of carbon: CO], pp. 222–223. In Photochemistry. New York: John Wiley & Sons, Inc., 1966.
68. Campbell, J. A. Growth, fertility etc. in animals during attempted acclimatization to carbon monoxide. Q. J. Exp. Physiol. 24:271–281, 1934.
69. Campbell, J. A. Hypertrophy of the heart in acclimatization to chronic carbon monoxide poisoning. J. Physiol. (Lond.) 77:8P–9P, 1933.
70. Campbell, J. A. Tissue oxygen tension and carbon monoxide poisoning. J. Physiol. (Lond.) 68:81–96, 1929.
71. Cavanagh, L. A., C. F. Schadt, and E. Robinson. Atmospheric hydrocarbon and carbon monoxide measurements at Point Barrow, Alaska. Environ. Sci. Technol. 3:251–257, 1969.
72. Cavender, J. H., D. S. Kircher, and A. J. Hoffman. Nationwide Air Pollutant Emission Trends, 1940–1970. Research Triangle Park, N.C.: U.S. Environmental Pro-

tection Agency, Office of Air and Water Programs, 1973. 52 pp.
73. Chance, B., M. Erecinska, and M. Wagner. Mitochondrial responses to carbon monoxide toxicity. Ann. N.Y. Acad. Sci. 174:193–204, 1970.
74. Chance, B., N. Oshino, T. Sugano, and A. Mayevsky. Basic principles of tissue oxygen determination from mitochondrial signals, pp. 277–292. In H. I. Bicher and D. F. Bruley, Eds. Oxygen Transport to Tissue. Instrumentation, Methods, and Physiology. Advances in Experimental Medicine and Biology. Vol. 37A. New York: Plenum Press, 1973.
75. Chanutin, A., and R. R. Curnish. Effect of organic and inorganic phospates on the oxygen equilibrium of human erythrocytes. Arch. Biochem. Biophys. 121:96–102, 1967.
76. Chapanis, A. The relevance of laboratory studies to practical situations. Ergonomics 10:557–577, 1967.
77. Chapanis, A. The search for relevance in applied research, pp. 1–14. In W. T. Singleton, J. G. Fox, and D. Whitfield, Eds. Measurement of Man at Work. An Appraisal of Physiological and Psychological Criteria in Man-Machine Systems. London: Taylor and Francis Limited, 1971.
78. Chevalier, R. B., R. A. Krumholz, and J. C. Ross. Reaction of nonsmokers to carbon monoxide inhalation. Cardiopulmonary responses at rest and during exercise. J.A.M.A. 198:1061–1064, 1966.
79. Chiodi, H., D. B. Dill, F. Consolazio, and S. M. Horvath. Respiratory and circulatory responses to acute carbon monoxide poisoning. Amer. J. Physiol. 134:683–693, 1941.
80. Chovin, P. Carbon monoxide: Analysis of exhaust gas investigations in Paris. Environ. Res. 1:198–216, 1967.
81. City of Chicago, Department of Environmental Control, Engineering Services Division. Chicago 1974 Emission Inventory Summary.
82. City of New York, Bureau of Technical Services, Department of Air Resources. Data Report Aerometric Network. Calendar Year 1974. 37 pp.
83. City of New York, Bureau of Technical Services, Department of Air Resources. Estimated Emission Inventory Summary. Technical Service Report 16. April 1976.
84. Clark, B. J., and R. F. Coburn. Mean myoglobin oxygen tension during exercise at maximal oxygen uptake. J. Appl. Physiol. 39:135–144, 1975.
85. Clark, R. T., Jr., and A. B. Otis. Comparative studies on acclimatization of mice to carbon monoxide and to low oxygen. Amer. J. Physiol. 169:285–294, 1952.
86. Clerc, M., and F. Barat. Cinétique des produits de décomposition de CO_2 par photolyse-éclair dans l'ultraviolet lointain. J. Chim. Phys. 63:1525–1529, 1966.
87. Clerc, M., and F. Barat. Kinetics of CO formation studies by far-uv flash photolysis of CO_2. J. Chem. Phys. 46:107–110, 1967.
88. Coburn, R. F. Enhancement by phenobarbital and diphenylhydantoin of carbon monoxide production in normal man. N. Engl. J. Med. 283:512–515, 1970.
89. Coburn, R. F. The carbon monoxide stores. Ann. N.Y. Acad. Sci. 174:11–22, 1970.
90. Coburn, R. F., G. K. Danielson, W. S. Blakemore, and R. E. Forster II. Carbon monoxide in blood: Analytical method and sources of error. J. Appl. Physiol. 19:510–515, 1964.
91. Coburn, R. F., R. E. Forster, and P. B. Kane. Considerations of the physiological variables that determine the blood carboxyhemoglobin concentration in man. J. Clin. Invest. 44:1899–1910, 1965.
92. Coburn, R. F., and P. B. Kane. Maximal erythrocyte and hemoglobin catabolism. J. Clin. Invest. 47:1435–1446, 1968.
93. Coburn, R. F., and L. B. Mayers. Myoglobin O_2 tension determined from measure-

ments of carboxymyoglobin in skeletal muscle. Amer. J. Physiol. 220:66–74, 1971.

94. Coburn, R. F., F. Ploegmakers, P. Gondrie, and F. Abboud. Myocardial myoglobin oxygen tension. Amer. J. Physiol. 224:870–876, 1973.
95. Coburn, R. F., W. J. Williams, and R. E. Forster. Effects of erythrocyte destruction on carbon monoxide production in man. J. Clin. Invest. 43:1098–1103, 1964.
96. Coburn, R. F., W. J. Williams, and S. B. Kahn. Endogenous carbon monoxide production in patients with hemolytic anemia. J. Clin. Invest. 45:460–468, 1966.
97. Cohen, N., and L. Heicklen. The oxidation of inorganic non-metallic compounds, pp. 1–137. In C. H. Bamford and C. F. H. Tipper, Eds. Comprehensive Chemical Kinetics. Vol. 6. Reactions of Non-Metallic Inorganic Compounds. New York: Elsevier Publishing Company, 1972.
98. Cohen, S. I., M. Deane, and J. R. Goldsmith. Carbon monoxide and survival from myocardial infarction. Arch. Environ. Health 19:510–517, 1969.
99. Cole, P. V., L. H. Hawkins, and D. Roberts. Smoking during pregnancy and its effects on the fetus. J. Obstet. Gynaecol. Brit. Commonw. 79:782–787, 1972.
100. Collins, V. G. Isolation, cultivation and maintenance of autotrophs, pp. 1–52. In J. R. Norris and D. W. Ribbons, Eds. Methods in Microbiology. Vol. 3B. New York: Academic Press, 1969.
101. Collison, H. A., F. L. Rodkey, and J. D. O'Neal. Determination of carbon monoxide in blood by gas chromatography. Clin. Chem. 14:162–171, 1968.
102. Colmant, H. J. Effect of low carbon monoxide concentrations on the sleep and waking pattern of the albino rat. Staub-Reinhalt. Luft (Engl. Ed.) 32(4):32–35, 1972.
103. Colucci, J. M., and C. R. Begeman. Carbon monoxide in Detroit, New York and Los Angeles air. Environ. Sci. Technol. 3:41–47, 1969.
104. Committee on Space Research (COSPAR), International Council of Scientific Unions. CIRA 1965. COSPAR International Reference Atmosphere 1965. Compiled by the Members of COSPAR Working Group IV, 1965. Amsterdam: North-Holland Publishing Company, 1965. 313 pp.
105. Comstock, G. W., and F. E. Lundin, Jr. Parental smoking and perinatal mortality. Amer. J. Obstet. Gynecol. 98:708–718, 1967.
106. Comstock, G. W., F. K. Shah, M. B. Meyer, and H. Abby. Low birth weight and neonatal mortality rate related to maternal smoking and socioeconomic status. Amer. J. Obstet. Gynecol. 111:53–59, 1971.
107. Conlee, C. J., P. A. Kenline, R. L. Cummins, and V. J. Konopinski. Motor vehicle exhausts at three selected sites. Arch. Environ. Health 14:429–446, 1967.
108. Connes, P., J. Connes, L. D. Kaplan, and W. S. Benedict. Carbon monoxide in the Venus atmosphere. Astrophys. J. 152:731–743, 1968.
109. Cooper, A. G. Carbon Monoxide. A Bibliography with Abstracts. Public Health Service Publication No. 1503. Washington, D.C.: U.S. Department of Health, Education and Welfare, 1966. 440 pp.
110. Coronary Drug Project Research Group. Factors infuencing long-term prognosis after recovery from myocardial infarction—three-year findings of the coronary drug project. J. Chron. Dis. 27:267–285, 1974.
111. Crespi, H. L., D. Huff, H. F. DaBoll, and J. J. Katz. Carbon Monoxide in the Biosphere: CO Emission by Fresh-Water Algae. Final Report CRC-APRAC-CAPA-4-68-5) to Coordinating Research Council and Air Pollution Control Office, Environmental Protection Agency. Argonne, Ill.: Argonne National Laboratory, 1972. 26 pp.
112. Crutzen, P. J. A discussion of the chemistry of some minor constituents in the stratosphere and troposphere. Pure Appl. Geophys. 106–108:1385–1399, 1973.

113. Crutzen, P. J. Photochemical reactions initiated by and influencing ozone in unpolluted tropospheric air. Tellus 26:47–57, 1974.
114. Curphey, T. J., L. P. L. Hood, and N. M. Perkins. Carboxyhemoglobin in relation to air pollution and smoking. Postmortem studies. Arch. Environ. Health 10:179–185, 1965.
115. Dahms, T. E., and S. M. Horvath. Rapid, accurate technique for determination of carbon monoxide in blood. Clin. Chem. 20:533–537, 1974.
116. Dahms, T. E., S. M. Horvath, and D. J. Gray. Technique for accurately producing desired carboxyhemoglobin levels during rest and exercise. J. Appl. Physiol. 38:366–368, 1975.
117. Davie, R., N. Butler, and H. Goldstein. From Birth to Seven. The Second Report of the National Child Development Study. (1958 Cohort) London: Longman, 1972. 198 pp.
118. DeBias, D. A., C. M. Banerjee, N. C. Birkhead, W. V. Harrer, and L. A. Kazal. Carbon monoxide inhalation effects following myocardial infarction in monkeys. Arch. Environ. Health 27:161–167, 1973.
119. deBruin, A. Carboxyhemoglobin levels due to traffic exhaust. Arch. Environ. Health 15:384–389, 1967.
120. Delwiche, C. C. Carbonic monoxide production and utilization by higher plants. Ann. N.Y. Acad. Sci. 174:116–121, 1970.
121. Denson, R., J. L. Nanson, and M. A. MacWatters. Hyperkinesis and maternal smoking. Can. Psychiatr. Assoc. J. 20:183–187, 1975.
122. Dinman, B. D., J. W. Eaton, and G. J. Brewer. Effects of carbon monoxide on DPG concentrations in the erythrocyte. Ann. N.Y. Acad. Sci. 174:246–251, 1970.
123. DiVincenzo, G. D., and M. L. Hamilton. Fate and disposition of ^{14}C methylene chloride in the rat. Toxicol. Appl. Pharmacol. 32:385–393, 1975.
124. Dixon-Lewis, G., W. E. Wilson, and A. A. Westenberg. Studies of hydroxyl radical kinetics by quantitative ERS. J. Chem. Phys. 44:2877–2884, 1966.
125. Dominquez, A. M., H. E. Christensen, L. R. Goldbaum, and V. A. Stembridge. A sensitive procedure for determining carbon monoxide in blood or tissue utilizing gas-solid chromatography. Toxicol. Appl. Pharmacol. 1:135–143, 1959.
126. Dorcus, R. M., and G. E. Weigand. The effect of exhaust gas on the performance in certain psychological tests. J. Gen. Psychol. 2:73–96, 1929.
127. Douglas, C. G., J. S. Haldane, and J. B. S. Haldane. The laws of combination of haemoglobin with carbon monoxide and oxygen. J. Physiol. (Lond.) 44:275–304, 1912.
128. Drabkin, D. L., and J. H. Austin. Spectrophotometric studies. II. Preparations from washed blood cells; nitric oxide hemoglobin and sulfhemoglobin. J. Biol. Chem. 112:51–65, 1935.
129. Drinkwater, B. L., P. B. Raven, S. M. Horvath, J. A. Gliner, R. O. Ruhling, N. W. Bolduan, and S. Taguchi. Air pollution, exercise, and heat stress. Arch. Environ. Health 28:177–181, 1974.
130. Dubois, L., and J. L. Monkman. Continuous determination of carbon monoxide by frontal analysis. Anal. Chem. 44:74–76, 1972.
131. Dubois, L., and J. L. Monkman. L'emploi de l'analyse frontale pour l'echantillonnage et le dosage de l'oxyde de carbone dans l'air. Mikrochim. Acta 1970:313–320.
132. Duvelleroy, M. A., H. Mehmel, and M. B. Laver. Hemoglobin-oxygen equilibrium and coronary blood flow: An analog model. J. Appl. Physiol. 35:480–484, 1973.
133. Eckardt, R. E., H. N. MacFarland, Y. C. E. Alarie, and W. H. Busey. The biologic effect from long-term exposure of primates to carbon monoxide. Arch. Environ. Health 25:381–387, 1972.

134. Ehrich, W. E., S. Bellet, and F. H. Lewey. Cardiac changes from CO poisoning. Amer. J. Med. Sci. 208:511–523, 1944.
135. Ekblom, B., and R. Huot. Response to submaximal and maximal exercise at different levels of carboxyhemoglobin. Acta Physiol. Scand. 86:474–482, 1972.
136. Essenberg, J. M., J. V. Schwind, and A. R. Patras. The effects of nicotine and cigarette smoke on pregnant female albino rats and their offsprings. J. Lab. Clin. Med. 25:708–717, 1940.
137. Estabrook, R. W., M. R. Franklin, and A. G. Hildebrandt. Factors influencing the inhibitory effect of carbon monoxide on cytochrome P-450-catalyzed mixed function oxidation reactions. Ann. N.Y. Acad. Sci. 174:218–232, 1970.
138. Faltings, K., W. Groth, and P. Harteck. Photochemische Untersuchungen im SCHUMANN-Ultraviolett Nr. 7. Zur Photochemie des Kohlenoxyds. Z. Physik. Chem. Abt. B. 41:15–22, 1938.
139. Fischer, E. R., and M. McCarty, Jr. Study of the reaction of electronically excited oxygen molecules with carbon monoxide. J. Chem. Phys. 45:781–784, 1966.
140. Flury, F., and F. Zernik. Kohlenoxyd, pp. 195–196. In Schädlich Gase. Dämpfe, Nebel, Rauch- und Staubarten. Berlin: Verlag von Julius Springer, 1931.
141. Fodor, G. G., and G. Winneke. Effect of low CO concentrations on resistance to monotony and on psychomotor capacity. Staub-Reinhalt. Luft (Engl. Ed.) 32(4): 46–54, 1972.
142. Forbes, W. H., D. B. Dill, H. DeSilva, and F. M. Van Deventer. The influence of moderate carbon monoxide poisoning upon the ability to drive automobiles. J. Ind. Hyg. Toxicol. 19:598–603, 1937.
143. Forbes, W. H., F. Sargent, and F. J. W. Roughton. The rate of carbon monoxide uptake by normal men. Amer. J. Physiol. 143:594–608, 1945.
144. Forster, R. E. Carbon monoxide and the partial pressure of oxygen in tissue. Ann. N.Y. Acad. Sci. 174:233–241, 1970.
145. Francisco, D. E., and J. K. G. Silvey. The effect of carbon monoxide inhibition on the growth of an aquatic streptomycete. Can. J. Microbiol. 17:347–351, 1971.
146. Frazier, T. M., G. H. Davis, H. Goldstein, and I. D. Goldberg. Cigarette smoking and prematurity: A prospective study. Amer. J. Obstet. Gynecol. 81:988–996, 1961.
147. Fristedt, B., and B. Akesson. Health hazards from automobile exhausts at service facilities of multistory garages. Hyg. Revy 60:112–118, 1971. (in Swedish)
148. Gaensler, E. A., J. B. Cadigan, Jr., M. F. Ellicott, R. H. Jones, and A. Marks. A new method for rapid precise determination of carbon monoxide in blood. J. Lab. Clin. Med. 49:945–957, 1957.
149. Gardner, R. A., and R. H. Petrucci. The chemisorption of carbon monoxide on metals. J. Amer. Chem. Soc. 82:5051–5053, 1960.
150. Garvin, D. The oxidation of carbon monoxide in the presence of ozone. J. Amer. Chem. Soc. 76:1523–1527, 1954.
151. General Electric Company. Indoor-Outdoor Carbon Monoxide Pollution Study. EPA-R4-73-020. Philadelphia: General Electric Company, 1972. [437 pp.]
152. Gennser, G., K. Maršál, and B. Brantmark. Maternal smoking and fetal breathing movements. Amer. J. Obstet. Gynecol. 123:861–867, 1975.
153. Gilbert, G. J., and G. H. Glaser. Neurologic manifestations of chronic carbon monoxide poisoning. N. Engl. J. Med. 261:1217–1220, 1959.
154. Gilson, R. D., F. E. Guedry, Jr., and A. J. Benson. Influence of vestibular stimulation and display luminance on the performance of a compensatory tracking task. Aerosp. Med. 41:1231–1237, 1970.
155. Ginsberg, M. D., and R. E. Myers. Fetal brain damages following maternal carbon monoxide intoxication: An experimental study. Acta Obstet. Gynecol. Scand. 53: 309–317, 1974.

156. Ginsberg, M. D., and R. E. Myers. Fetal brain injury after maternal carbon monoxide intoxication: Clinical and neuropathologic aspects. Neurology 26:15–23, 1976.
157. Glasson, W. A. Effect of carbon monoxide on atmospheric photooxidation of nitric oxide-hydrocarbon mixtures. Environ. Sci. Technol. 9:343–347, 1975.
158. Gliner, J. A., P. B. Raven, S. M. Horvath, B. L. Drinkwater, and J. C. Sutton. Man's physiologic response to long-term work during thermal and pollutant stress. J. Appl. Physiol. 39:628–632, 1975.
159. Gold, R. E., and L. L. Kulak. Effect of hypoxia on aircraft pilot performance. Aerosp. Med. 43:180–183, 1972.
160. Goldman, A., D. G. Murcray, F. H. Murcray, W. J. Williams, J. N. Brooks, and C. M. Bradford. Vertical distribution of CO in the atmosphere. J. Geophys. Res. 78:5273–5283, 1973.
161. Goldsmith, J. R., and S. A. Landaw. Carbon monoxide and human health. Science 162:1352–1359, 1968.
162. Goldstein, H. Cigarette smoking and low-birthweight babies. Amer. J. Obstet. Gynecol. 114:570–571, 1972. (correspondence)
163. Goldstein, H. Factors influencing the height of seven year old children—results from the National Child Development Study. Human Biol. 43:92–111, 1971.
164. Gorbatow, O., and L. Noro. On acclimatization in connection with acute carbon monoxide poisonings. Acta Physiol. Scand. 15:77–87, 1948.
165. Gordon, T. Heart Disease in Adults, United States—1960–1962. U.S. Department of Health, Education, and Welfare, Public Health Service, National Center for Health Statistics. Vital and Health Statistics Series 11, No. 6. Washington, D.C.: U.S. Government Printing Office, 1964. 43 pp.
166. Gordon, T., W. B. Kannel, D. McGee, and T. R. Dawber. Death and coronary attacks in men after giving up cigarette smoking. Lancet 2:1345–1348, 1974.
167. Gorse, R. A., and D. H. Volman. Photochemistry of the gaseous hydrogen peroxide–carbon monoxide system: Rate constants for hydroxyl radical reactions with hydrogen peroxide and isobutane by competitive kinetics. J. Photochem. 1:1–10, 1972.
168. Göthert, M., F. Lutz, and G. Malorny. Carbon monoxide partial pressure in tissue of different animals. Environ. Res. 3:303–309, 1970.
169. Goujard, J., C. Rumeau, and D. Schwartz. Smoking during pregnancy, stillbirth and abruptio placentae. Biomedicine 23:20–22, 1975.
170. Grahn, D., and J. Kratchman. Variations in neonatal death rate and birth weight in the United States and possible relations to environmental radiation, geology, and altitude. Amer. J. Hum. Genet. 15:329–352, 1963.
171. Graven, W. M., and F. J. Long. Kinetics and mechanisms of the two opposing reactions of the equilibrium $CO + H_2O = CO_2 + H_2$. J. Amer. Chem. Soc. 76: 2602–2607, 1954.
172. Green, A. E. S., T. Sawada, B. C. Edgar, and M. A. Uman. Production of carbon monoxide by charged particle deposition. J. Geophys. Res. 78:5284–5292, 1973.
173. Gregg, D. E., and L. C. Fisher. [Blood supply to the heart], p. 1547. In W. F. Hamilton and P. Dow, Eds. Handbook of Physiology. Sect. 2, Vol. 2, Circulation. Washington, D.C.: American Physiological Society, 1963.
174. Greiner, N. R. Hydroxyl-radical kinetics by kinetic spectroscopy. I. Reactions with H_2, CO, and CH_4 at 300°K. J. Chem. Phys. 46:2795–2799, 1967.
175. Groll-Knapp, E., H. Wagner, H. Hauck, and M. Haider. Effects of low carbon monoxide concentrations on vigilance and computer-analyzed brain potentials. Staub-Reinhalt. Luft (Engl. Ed.) 32(4):64–68, 1972.
176. Groth, W., W. Pessara, and H. J. Rommel. Photochemische Untersuchungen im SCHUMANN-Ultraviolett Nr. 11. Die photochemische Zersetzung von N_2 und CO

im Lichte der Xenon- und Krypton-Resonanzwellenlängen. Z. Physik. Chem. (Frankfurt) 32:192–211, 1962.

177. Grut, A. Chronic Carbon Monoxide Poisoning. A Study in Occupational Medicine. Copenhagen: Ejnar Munksgaard, 1949. 229 pp.

178. Guest, A. D. L., C. Duncan, and P. J. Lawther. Carbon monoxide and phenobarbitone: A comparison of effects on auditory flutter fusion threshold and critical flicker fusion threshold. Ergonomics 13:587–594, 1970.

179. Haddon, W., Jr., R. E. L. Nesbitt, and R. Garcia. Smoking and pregnancy: Carbon monoxide in blood during gestation and at term. Obstet. Gynecol. 18:262–267, 1961.

180. Haebisch, H. Die Zigarette als Kohlenmonoxydquelle. Arch. Toxicol. 26:251–261, 1970.

181. Haggard, H. W. Studies in carbon monoxide asphyxia. I. The behavior of the heart. Amer. J. Physiol. 56:390–403, 1921.

182. Haider, M., E. Groll-Knapp, H. Höller, M. Neuberger, and H. Stidl. Effects of moderate CO dose on the central nervous system—electrophysiological and behavior data and clinical relevance, pp. 217–232. In A. J. Finkel and W. C. Duel, Eds. Clinical Implications of Air Pollution Research. American Medical Association Air Pollution Medical Research Conference, 1974. Acton, Mass.: Publishing Science Group, Inc., 1976.

183. Haldane, J. The action of carbonic oxide on man. J. Physiol. (Lond.) 18:430–462, 1895.

184. Halperin, M. H., R. A. McFarland, J. I. Niven, and F. J. W. Roughton. The time course of the effects of carbon monoxide on visual thresholds. J. Physiol. (Lond.) 146:583–593, 1959.

185. Hampson, R. F., Jr., and D. Garvin, Eds. Chemical Kinetic and Photochemical Data for Modelling Atmospheric Chemistry. NBS Technical Note 866. Washington, D.C.: U.S. Department of Commerce, National Bureau of Standards, 1975. 113 pp.

186. Hanks, T. G. Human performance of a psychomotor test as a function of exposure to carbon monoxide. Ann. N.Y. Acad. Sci. 174:421–424, 1970.

187. Harteck, P., and S. Dondes. Reaction of carbon monoxide and ozone. J. Chem. Phys. 26:1734–1737, 1957.

188. Heicklen, J. Gas phase oxidation of perhalocarbons. Adv. Photochem. 7:57–148, 1969.

189. Heicklen, J., K. Westberg, and N. Cohen. The Conversion of NO to NO_2 in Polluted Atmospheres. Center for Air Environment Studies Publ. 115–169. University Park: The Pennsylvania State University, 1969. 5 pp.

190. Hellman, L. M., and J. A. Pritchard. Williams Obstetrics. (14th ed.) New York: Appleton-Century-Crofts, 1971. 1,242 pp.

191. Heron, H. J. The effects of smoking during pregnancy: A review with a preview. N. Z. Med. J. 61:545–548, 1962.

192. Herzberg, G. Molecular Spectra and Molecular Structure. I. Spectra of Diatomic Molecules. (2nd ed.) New York: D. Van Nostrand Company, Inc., 1950. 658 pp.

193. Heslop-Harrison, J., and Y. Heslop-Harrison. Studies on flowering-plant growth and organogenesis. II. The modification of sex expression in *Cannabis sativa* by carbon monoxide. Proc. Roy. Soc. Edinburgh B 66:424–434, 1957.

194. Hesstvedt, E. Vertical distribution of CO near the tropopause. Nature 225:50, 1970.

195. Hill, E. P., J. R. Hill, G. G. Power, and L. D. Longo. Carbon monoxide exchanges between the human fetus and mother: A mathematical model. Amer. J. Physiol. 232:H311–H323, 1977.

196. Honig, C. R., and J. Bourdeau-Martini. O_2 and the number and arrangement of

coronary capillaries; effect on calculated tissue PO_2, pp. 519–524. In H. I. Bicher and D. F. Bruley, Eds. Oxygen Transport to Tissue. Instrumentation, Methods, and Physiology. Advances in Experimental Medicine and Biology. Vol. 37A. New York: Plenum Press, 1973.

197. Hopfield, J. J., and R. T. Birge. Ultra-violet absorption and emission spectra of carbon monoxide. Phys. Rev. 29:922, 1927. (abstract)

198. Horvath, M., and E. Frantik. Quantitative interpretation of experimental toxicological data: The use of reference substances, pp. 11–21. In M. Horvath, Ed. Adverse Effects of Environmental Chemicals and Psychotropic Drugs. Vol. 1. Amsterdam: Elsevier Scientific Publishing Company, 1973.

199. Horvath, S. M., T. E. Dahms, and J. F. O'Hanlon. Carbon monoxide and human vigilance. A deleterious effect of present urban concentrations. Arch. Environ. Health 23:343–347, 1971.

200. Horvath, S. M., P. B. Raven, T. E. Dahms, and D. J. Gray. Maximal aerobic capacity at different levels of carboxyhemoglobin. J. Appl. Physiol. 38:300–303, 1975.

201. Hosko, M. J. The effect of carbon monoxide on the visual evoked response in man and the spontaneous electroencephalogram. Arch. Environ. Health 21:174–180, 1970.

202. Howlett, L., and R. J. Shephard. Carbon monoxide as a hazard in aviation. J. Occup. Med. 15:874–877, 1973.

203. Hull, H. M., and F. W. Went. Life processes of plants as affected by air pollution, pp. 122–128. In Proceedings of the Second National Air Pollution Symposium, Pasadena, California, 1952. Los Angeles: National Air Pollution Symposium, 1952.

204. Hwang, J. C., C. H. Chen, and R. H. Burris. Inhibition of nitrogenase-catalyzed reductions. Biochim. Biophys. Acta 292:256–270, 1973.

205. Ingersoll, R. B. The Capacity of the Soil as a Natural Sink for Carbon Monoxide. Final Report. SRI Project LSU-1380. Menlo Park, Calif.: Stanford Research Institute, 1972. 38 pp.

206. Ingersoll, R. B., R. E. Inman, and W. R. Fisher. Soil's potential as a sink for atmospheric carbon monoxide. Tellus 26:151–159, 1974.

207. Inman, R. E., and R. B. Ingersoll. Uptake of carbon monoxide by soil fungi. J. Air Pollut. Control Assoc. 21:646–647, 1971.

208. Inman, R. E., R. B. Ingersoll, and E. A. Levy. Soil: A natural sink for carbon monoxide. Science 172:1229–1231, 1971.

209. Jaeger, J. J., and J. J. McGrath. Effects of hypothermia on heart and respiratory responses of chick to carbon monoxide. J. Appl. Physiol. 34:564–567, 1973.

210. Jaffe, L. S. Ambient carbon monoxide and its fate in the atmosphere. J. Air Pollut. Control Assoc. 18:534–540, 1968.

211. Jaffe, L. S. Carbon monoxide in the biosphere: Sources, distributions, and concentrations. J. Geophys. Res. 78:5293–5305, 1973.

212. Jenkins, C. E. The haemoglobin concentration of workers connected with internal combustion engines. J. Hyg. 32:406–408, 1932.

213. Joels, N., and L. G. C. Pugh. The carbon monoxide dissociation curve of human blood. J. Physiol. (Lond.) 142:63–77, 1958.

214. Johnson, B. L., H. H. Cohen, R. Struble, J. V. Setzer, W. K. Anger, B. D. Gutnik, T. McDonough, and P. Hauser. Field evaluation of carbon monoxide exposed toll collectors, pp. 306–328. In C. Xintaras, B. L. Johnson, and I. de Groot, Eds. Behavioral Toxicology: Early Detection of Occupational Hazards. HEW Publ. No. (NIOSH) 74–126. Washington, D.C.: U.S. Department of Health, Education, and Welfare, 1974.

215. Johnson, K. L., L. H. Dworetzky, and A. N. Heller. Carbon Monoxide and Air Pollution from Automobile Emissions in New York City. New York: New York City Department of Air Pollution Control, 1967. 33 pp.
216. Johnson, K. L., L. H. Dworetzky, and A. N. Heller. Carbon monoxide and air pollution from automobile emissions in New York City. Science 160:67–68, 1968.
217. Jones, R. H., M. F. Ellicott, J. B. Cadigan, and E. A. Gaensler. The relationship between alveolar and blood carbon monoxide concentrations during breathholding. Simple estimation of COHb saturation. J. Lab. Clin. Med. 51:553–564, 1958.
218. Junge, C. E. Air Chemistry and Radioactivity. New York: Academic Press, Inc., 1963. 382 pp.
219. Junge, C. E., W. Seiler, and P. Warneck. The atmospheric ^{12}CO and ^{14}CO budget. J. Geophys. Res. 76:2866–2879, 1971.
220. Kagan, A., B. R. Harris, W. Winkelstein, Jr., K. G. Johnson, H. Kato, S. L. Syme, G. G. Rhoads, M. L. Gay, M. Z. Nichaman, H. B. Hamilton, and J. Tillotson. Epidemiologic studies of coronary heart disease and stroke in Japanese men living in Japan, Hawaii and California: Demographic, physical, dietary and biochemical characteristics. J. Chron. Dis. 27:345–364, 1974.
221. Kahn, A., R. B. Rutledge, D. L. Davis, J. A. Altes, G. E. Gantner, C. A. Thornton, and N. D. Wallace. Carboxyhemoglobin sources in the metropolitan St. Louis population. Arch. Environ. Health 29:127–135, 1974.
222. Kaplan, L. D., J. Connes, and P. Connes. Carbon monoxide in the Martian atmosphere. Astrophys. J. 157:L187–L192, 1969. (letter)
223. Kennedy, A. C., and D. J. Valtis. The oxygen dissociation curve in anemia of various types. J. Clin. Invest. 33:1372–1381, 1954.
224. Kessler, M., H. Lang, E. Sinagowitz, R. Rink, and J. Höper. Homeostasis of oxygen supply in liver and kidney, pp. 351–360. In H. I. Bicher and D. F. Bruley. Oxygen Transport to Tissue. Instrumentation, Methods, and Physiology. Advances in Experimental Medicine and Biology. Vol. 37A. New York: Plenum Press, 1973.
225. Keys, A., Ed. Coronary Heart Disease in Seven Countries. American Heart Association Monograph Number 29. Circulation 41(Suppl. I):I-1–I-211, 1970.
226. Killick, E. M. The acclimatization of mice to atmospheres containing low concentrations of carbon monoxide. J. Physiol. (Lond.) 91:279–292, 1937.
227. Killick, E. M. The acclimatization of the human subject to atmospheres containing low concentrations of carbon monoxide. J. Physiol. (Lond.) 87:41–55, 1936.
228. Killick, E. M. The nature of the acclimatization occurring during repeated exposure of the human subject to atmospheres containing low concentrations of carbon monoxide. J. Physiol. (Lond.) 107:27–44, 1948.
229. Kjeldsen, K., P. Astrup, and J. Wanstrup. Ultrastructural intimal changes in the rabbit aorta after a moderate carbon monoxide exposure. Atherosclerosis 16:67–82, 1972.
230. Kjeldsen, K., and F. Damgaard. Influence of prolonged carbon monoxide exposure and high altitude on the composition of blood and urine in man. Scand. J. Clin. Lab. Invest. 22(Suppl. 103):20–25, 1968.
231. Kjeldsen, K., H. K. Thomsen, and P. Astrup. Effects of carbon monoxide on myocardium. Ultrastructural changes in rabbits after moderate, chronic exposure. Circ. Res. 34:339–348, 1974.
232. Klausen, K., B. Rasmussen, H. Gjellerod, H. Madsen, and E. Petersen. Circulation, metabolism and ventilation during prolonged exposure to carbon monoxide and to high altitude. Scand. J. Clin. Lab. Invest. 22(Suppl. 103):26–38, 1968.
233. Klebba, A. J., J. D. Maurer, and E. J. Glass. Mortality Trends for Leading Causes of Death, United States—1950–1969. U.S. Department of Health, Education, and

Welfare, National Center for Health Statistics. Vital and Health Statistics Series 20, No. 16. Washington, D.C.: U. S. Government Printing Office, 1974. 75 pp.
234. Klendshoj, N. C., M. Feldstein, and A. L. Sprague. The spectrophotometric determination of carbon monoxide. J. Biol. Chem. 183:297–303, 1950.
235. Kluyver, A. J., and C. G. T. Schnellen. On the fermentation of carbon monoxide by pure cultures of methane bacteria. Arch. Biochem. 14:57–70, 1947.
236. Koob, G. F., Z. Annau, R. J. Rubin, and M. R. Montgomery. Effect of hypoxic hypoxia and carbon monoxide on food intake, water intake, and body weight in two strains of rats. Life Sci. 14:1511–1520, 1974.
237. Kortschak, H. P., and L. G. Nickell. Photosynthetic carbon monoxide metabolism by sugarcane leaves. Plant Sci. Lett. 1:213–216, 1973.
238. Knight, L. I., and W. Crocker. Toxicity of smoke. Contributions from the Hull Botanical Laboratory 171. Bot. Gazette 55:337–371, 1913.
239. Krall, A. R., and N. E. Tolbert. A comparison of the light dependent metabolism of carbon monoxide by barley leaves with that of formaldehyde, formate and carbon dioxide. Plant Physiol. 32:321–326, 1957.
240. Krause, A. Mechanismus der katalytischen Oxydation von CO mit N_2O. Bull. Acad. Polon. Sci. Ser. Sci. Chim. 9:5–6, 1961.
241. Kubic, V. L., M. W. Anders, R. R. Engel, C. H. Barlow, and W. S. Caughey, Metabolism of dihalomethanes to carbon monoxide. 1. *In vivo* studies. Drug. Metab. Dispos. 2:53–57, 1974.
242. Kullander, S., and B. Källen. A prospective study of smoking and pregnancy. Acta Obstet. Gynecol. Scand. 50:83–94, 1971.
243. Kuller, L., J. Perper, and M. Cooper. Sudden and unexplained death due to arteriosclerotic heart disease, pp. 292–332. In M. F. Oliver, Ed. Modern Trends in Cardiology. Vol. 3. London: Butterworths, 1975.
244. Kuller, L. H., E. P. Radford, D. Swift, J. A. Perper, and R. Fisher. Carbon monoxide and heart attacks. Arch. Environ. Health 30:477–482, 1975.
245. Kummler, R. H., and T. Baurer. A temporal model of tropospheric carbon-hydrogen chemistry. J. Geophys. Res. 78:5306–5316, 1973.
246. Kummler, R. H., R. N. Grenda, T. Baurer, M. H. Bortner, J. H. Davies, and J. MacDowall. Satellite solution of the carbon monoxide sink anomaly. EOS Trans. Amer. Geophys. Union 50(4):174, 1969. (abstract)
247. Lambert, J. L., R. R. Tschorn, and P. A. Hamlin. Determination of carbon monoxide in blood. Anal. Chem. 44:1529–1530, 1972.
248. Lamontagne, R. A., J. W. Swinnerton, and V. J. Linnenbom. Nonequilibrium of carbon monoxide and methane at the air-sea interface. J. Geophys. Res. 76:5117–5121, 1971.
249. Larsen, R. I., and H. Burke. Ambient Carbon Monoxide Exposure. Paper 69–167 Presented at 62nd Annual Meeting of the Air Pollution Control Association, New York, 1969. 39 pp.
250. Laties, V. G. On the use of reference substances in behavioral toxicology, pp. 83–88. In M. Horvath, Ed. Adverse Effects of Environmental Chemicals and Psychotropic Drugs. Vol. 1. Amsterdam: Elsevier Scientific Publishing Company, 1973.
251. Laties, V. G., and B. Weiss. Performance enhancement by the amphetamines: A new appraisal, pp. 800–808. In H. Brill, J. O. Cole, P. Deniker, H. Hippius, and P. B. Bradley, Eds. Neuropsycho-Pharmacology. Proceedings of the Fifth International Congress of the Collegium Internationale Neuro-Psycho-Pharmacologicum. Washington, D.C., 28–31 March, 1966. International Congress Series No. 129. Amsterdam: Excerpta Medica Foundation, 1967.

252. Leighton, P. A. Other reactions of oxygen atoms, pp. 146–150. In Photochemistry of Air Pollution. New York: Academic Press, 1961.
253. Lenfant, C., J. Torrance, E. English, C. A. Finch, C. Reynafarje, J. Ramos, and J. Faura. Effect of altitude on oxygen binding by hemoglobin and on organic phosphate levels. J. Clin. Invest. 47:2652–2656, 1968.
254. Levy, H., II. Normal atmosphere: Large radical and formaldehyde concentrations predicted. Science 173:141–143, 1971.
255. Levy, H., II. Photochemistry of minor constituents in the troposphere. Planet. Space Sci. 21:575–591, 1973.
256. Levy, H., II. Photochemistry of the lower troposphere. Planet. Space Sci. 20:919–935, 1972.
257. Lewey, F. H., and D. L. Drabkin. Experimental chronic carbon monoxide poisoning of dogs. Amer. J. Med. Sci. 208:502–511, 1944.
258. Lewin, L. Die Kohlenoxydvergiftung. Ein Handbuch für Mediziner, Techniker und Unfallrichter. Berlin: Julius Springer, 1920. 370 pp.
259. Lewis, J., A. D. Baddeley, K. G. Bonham, and D. Lovett. Traffic pollution and mental efficiency. Nature 225:95–97, 1970.
260. Liberti, A. The nature of particulate matter. Pure Appl. Chem. 24:631–642, 1970.
261. Lichty, J. A., R. Y. Ting, P. D. Bruns, and E. Dyar. Studies of babies born at high altitudes. I. Relation of altitude to birth weight. A.M.A. J. Dis. Child. 93: 666–669, 1957.
262. Lighthart, B. Survival of airborne bacteria in a high urban concentration of carbon monoxide. Appl. Microbiol. 25:86–91, 1973.
263. Lilienthal, J. L., Jr. Carbon monoxide. Pharmacol. Rev. 2:234–254, 1950.
264. Lilienthal, J. L., Jr., and C. H. Fugitt. The effect of low concentrations of carboxyhemoglobin on the "altitude tolerance" of man. Amer. J. Physiol. 145:359–364, 1946.
265. Lind, C. J., and P. W. Wilson. Carbon monoxide inhibition of nitrogen fixation by azotobacter. Arch. Biochem. 1:59–72, 1942.
266. Lind, C. J., and P. W. Wilson. Mechanism of biological nitrogen fixation. VIII. Carbon monoxide as an inhibitor for nitrogen fixation by red clover. J. Amer. Chem. Soc. 63:3511–3514, 1941.
267. Linderholm, H., T. Sjöstrand, and B. Söderström. A method for determination of low carbon monoxide concentration in blood. Acta Physiol. Scand. 66:1–8, 1966.
268. Link, W. T., E. A. McClatchie, D. A. Watson, and A. S. Compher. A Fluorescent Source NDIR Carbon Monoxide Analyzer. Presented at Conference on Methods in Air Pollution and Industrial Hygiene Studies, 12th, Los Angeles, California, April 6–8, 1971.
269. Linnenbom, J. V., J. W. Swinnerton, and R. A. Lamontagne. The ocean as a source for atmospheric carbon monoxide. J. Geophys. Res. 78:5333–5340, 1973.
270. Liss, P. S., and P. G. Slater. Flux of gases across the air–sea interface. Nature 247:181–184, 1974.
271. Lissi, E., R. Simonaitis, and J. Heicklen. The bromine atom catalyzed oxidation of carbon monoxide. J. Phys. Chem. 76:1416–1419, 1972.
272. Locke, J. L., and L. Herzberg. The absorption due to carbon monoxide in the infrared solar sprectrum. Can. J. Phys. 31:504–516, 1953.
273. Lockshin, A., and R. H. Burris. Inhibitors of nitrogen fixation in extracts from *Clostridium pasteurianum*. Biochim. Biophys. Acta 111:1–10, 1965.
274. Longo, L. D. Carbon monoxide: Effects on oxygenation of the fetus in utero. Science 194:523–525, 1976.

275. Longo, L. D. Carbon monoxide in the pregnant mother and fetus and its exchange across the placenta. Ann. N.Y. Acad. Sci. 174:313–341, 1970.
276. Longo, L. D. The biologic effects of carbon monoxide on the pregnant woman, the fetus and newborn infant. Amer. J. Obstet. Gynecol. (in press)
277. Longo, L. D., and K. Ching. Placental diffusion capacity for carbon monoxide and oxygen in unanesthetized sheep. Amer. J. Physiol. (in press)
278. Longo, L. D., and E. P. Hill. Carbon monoxide uptake and elimination in fetal and maternal sheep. Amer. J. Physiol. 232:H324–H330, 1977.
279. Longo, L. D., G. G. Power, and R. E. Forster, II. Respiratory function of the placenta as determined with carbon monoxide in sheep and dogs. J. Clin. Invest. 46:812–828, 1967.
280. Los Angeles County Air Pollution Control District. 1974 Profile of Air Pollution Control. Los Angeles: County of Los Angeles, Air Pollution Control District, 1974. 91 pp.
281. Lowe, C. R. Effect of mothers' smoking habits on birthweight of their children. Brit. Med. J. 2:673–676, 1959.
282. Luomanmäki, K. Studies on the metabolism of carbon monoxide. Ann. Med. Exp. Biol. Fenniae 44(Suppl. 2):1–55, 1966.
283. Mackworth, J. F. Vigilance and Attention. Baltimore: Penguin Books, Inc., 1970. 188 pp.
284. Mackworth, J. F. Vigilance and Habituation. Baltimore: Penguin Books, Inc., 1969. 237 pp.
285. Mackworth, N. H. Researches on the Measurement of Human Performance. Medical Research Council Special Report Series No. 268. London: His Majesty's Stationery Office, 1950. 156 pp.
286. Madley, D. G., and R. F. Strickland-Constable. The kinetics of the oxidation of charcoal with nitrous oxide. Trans. Faraday Soc. 49:1312–1324, 1953.
287. Malenfant, A. L., S. R. Gambino, A. J. Waraksa, and E. I. Roe. Spectrophotometric determination of hemoglobin concentration and per cent oxyhemoglobin and carboxyhemoglobin saturation. Clin. Chem. 14:789, 1968. (abstract)
288. Manning, F., E. Wyn Pugh, and K. Boddy. Effects of cigarette smoking on fetal breathing movements in normal pregnancies. Brit. Med. J. 1:552–553, 1975.
289. Manning, F. A., and C. Feyerabend. Cigarette smoking and fetal breathing movements. Brit. J. Obstet. Gynaecol. 83:262–270, 1976.
290. Maxwell, J. C., C. H. Barlow, J. E. Spallholz, and W. S. Caughey. The utility of infrared spectroscopy as a probe of intact tissue: Determination of carbon monoxide and hemeproteins in blood and heart muscle. Biochem. Biophys. Res. Commun. 61:230–236, 1974.
291. McClatchie, E. A. Development of an Infrared Fluorescent Gas Analyzer. EPA-R2-72-121. Berkeley, Calif.: Akron Scientific Labs, 1972. 7 pp.
292. McClatchie, E. A., A. B. Compher, and K. G. Williams. A high specificity carbon monoxide analyzer. Anal. Instrum. (Symposium) 10:67–69, 1972.
293. McConnell, J. C., M. B. McElroy, and S. C. Wofsy. Natural sources of atmospheric CO. Nature 233:187–188, 1971.
294. McCormick, R. A., and C. Xintaras. Variation of carbon monoxide concentrations as related to sampling interval, traffic and meteorological factors. J. Appl. Meteorol. 1:237–243, 1962.
295. McCullough, J. D., R. A. Crane, and A. O. Beckman. Determination of carbon monoxide in air by use of red mercuric oxide. Ind. Eng. Chem. Anal. Ed. 19:999–1002, 1947.

296. McFarland, R. A. Low level exposure to carbon monoxide and driving performance. Arch. Environ. Health 27:355–359, 1973.
297. McFarland, R. A. The effects of exposure to small quantities of carbon monoxide on vision. Ann. N.Y. Acad. Sci. 174:301–312, 1970.
298. McFarland, R. A., W. H. Forbes, H. W. Stoudt, J. D. Dougherty, T. J. Crowley, R. C. Moore, and T. J. Nalwalk. A Study of the Effects of Low Levels of Carbon Monoxide upon Humans Performing Driving Tasks. Final Report, CRC-APRAC Contract CAPM-9-69(2–70), June 15, 1970–September 15, 1972. Boston: Guggenheim Center for Aerospace Health and Safety, Harvard School of Public Health, 1973. 88 pp.
299. McFarland, R. A., W. H. Forbes, H. W. Stoudt, J. D. Dougherty, A. J. Morandi, and T. J. Nalwalk. A Study of the Effects of Low Levels of Carbon Monoxide upon Humans Performing Driving Tasks. Final Report. June 15, 1970–June 14, 1971. Boston: Guggenheim Center for Aerospace Health and Safety, Harvard School of Public Health, 1971. 56 pp.
300. McFarland, R. A., D. J. W. Roughton, M. H. Halperin, and J. I. Niven. The effects of carbon monoxide and altitude on visual thresholds. J. Aviat. Med. 15: 381–394, 1944.
301. McGrath, J. J., and J. Jaeger. Effect of iodoacetate on the carbon monoxide tolerance of the chick. Respir. Physiol. 12:71–76, 1971.
302. McKee, H. C., and R. E. Childers. Collaborative Study of Reference Method for the Continuous Measurement of Carbon Monoxide in the Atmosphere (Non-Dispersive Infrared Spectrometry). EPA-R4-72-009. Houston, Texas: Southwest Research Institute, 1972. [41 pp.]
303. McMillan, C., and J. M. Cope. Response to carbon monoxide by geographic variants in *Acacia farnesiana.* Amer. J. Bot. 56:600–602, 1969.
304. McMullen, T. B. Interpreting the eight-hour national ambient air quality standard for carbon monoxide. J. Air Pollut. Control Assoc. 25:1009–1014, 1975.
305. McNesby, J. R., and H. Okabe. Vacuum ultraviolet photochemistry. Adv. Photochem. 3:157–240, 1964.
306. Mellor, J. W. The chemical properties of carbon monoxide, pp. 926–950. In A Comprehensive Treatise on Inorganic and Theoretical Chemistry. Vol. V. B, Al, Ga, In, Tl, Sc, Ce and Rare Earth Metals, C (Part 1). London: Longmans, Green and Co., 1924.
307. Menser, H. A., J. J. Grosso, H. E. Heggstad, and O. E. Street. Air filtration study of "hidden" air pollution injury to tobacco plants. Plant Physiol. 39(Suppl.):lviii, 1964. (abstract)
308. Meyer, M. B., and G. W. Comstock. Maternal cigarette smoking and perinatal mortality. Amer. J. Epidemiol. 96:1–10, 1972.
309. Meyer, M. B., B. S. Jonas, and J. A. Tonascia. Perinatal events associated with maternal smoking during pregnancy. Amer. J. Epidemiol. 103:464–476, 1976.
310. Meyer, M. B., J. A. Tonascia, and C. Buck. The interrelationship of maternal smoking and increased perinatal mortality with other risk factors. Further analysis of the Ontario perinatal mortality study, 1960–1961. Amer. J. Epidemiol. 100:443–452, 1974.
311. Migeotte, M. The fundamental band of carbon monoxide at 4.7 μ in the solar spectrum. Phys. Rev. 75:1108–1109, 1949.
312. Migeotte, M., and L. Neven. Recents proges dans l'observation du spectre infrarouge du soleil à la station scientifique du Jungfraujoch (Suisse). Mem. Soc. Roy. Sci. Liege 12:165–178, 1952.
313. Mikulka, P., R. O'Donnell, P. Heinig, and J. Theodore. The effect of carbon mon-

oxide on human performance. Ann. N.Y. Acad. Sci. 174:409–420, 1970.

314. Minina, E. G., and L. G. Tylkina. Physiological study of the effect of gases upon sex differentiation in plants. Dokl. Akad. Nauk. SSSR 55(2):165–168, 1947. (in Russian)
315. Miranda, J. M., V. J. Konopinski, and R. I. Larsen. Carbon monoxide control in a high highway tunnel. Arch. Environ. Health 15:16–25, 1967.
316. Moeller, T. Carbon monoxide, pp. 685–687. In Inorganic Chemistry. An Advanced Textbook. New York: John Wiley & Sons, Inc., 1952.
317. Montgomery, M. R., and R. J. Rubin. Adaptation to the inhibitory effect of carbon monoxide inhalation on drug metabolism. J. Appl. Physiol. 35:601–607, 1973.
318. Montgomery, M. R., and R. J. Rubin. Oxygenation during inhalation of drug metabolism by carbon monoxide or hypoxic hypoxia. J. Appl. Physiol. 35:505–509, 1973.
319. Montgomery, M. R., and R. J. Rubin. The effect of carbon monoxide inhalation on *in vivo* drug metabolism in the rat. J. Pharmacol. Exp. Ther. 179:465–473, 1971.
320. Mourant, R. R., and T. H. Rockwell. Strategies of visual search by novice and experienced drivers. Human Factors 14:325–335, 1972.
321. Mulhausen, R. O., P. Astrup, and K. Mellemgaard. Oxygen affinity and acid-base status of human blood during exposure to hypoxia and carbon monoxide. Scand. J. Clin. Lab. Invest. 22(Suppl. 103):9–15, 1968.
322. Müller, R. H. A supersensitive gas detector permits accurate detection of toxic or combustible gases in extremely low concentrations. Anal. Chem. 26(9):39A–42A, 1954.
323. Murray, J. F., P. Gold, and B. L. Johnson, Jr. Systemic oxygen transport in induced normovolemic anemia and polycythemia. Amer. J. Physiol. 203:720–724, 1962.
324. Musselman, N. P., W. A. Groff, P. P. Yevich, F. T. Wilinski, M. H. Weeks, and F. W. Oberst. Continuous exposure of laboratory animals to low concentrations of carbon monoxide. Aerosp. Med. 30:524–529, 1959.
325. Nasmith, G. G., and D. A. L. Graham. The haematology of carbon-monoxide poisoning. J. Physiol. (Lond.) 35:32–52, 1906.
326. National Research Council. Division of Medical Sciences. Committee on Effects of Atmospheric Contaminants on Human Health and Welfare. Effects of Chronic Exposure to Low Levels of Carbon Monoxide on Human Health, Behavior, and Performance. Washington, D.C.: National Academy of Sciences, 1969. 66 pp.
327. Neill, W. A. Effects of arterial hypoxemia and hyperoxia on oxygen availability for myocardial metabolism. Patients with and without coronary heart disease. Amer. J. Cardiol. 24:166–171, 1969.
328. Nicholson, S. E. A pollution model for street-level air. Atmos. Environ. 9:19–31, 1975.
329. Nielsen, B. Thermoregulation during work in carbon monoxide poisoning. Acta Physiol. Scand. 82:98–106, 1971.
330. Niswander, K. R., and M. Gordon. Cigarette smoking, pp. 72–80. In The Women and Their Pregnancies. The Collaborative Perinatal Study of the National Institute of Neurological Diseases and Stroke. Philadelphia: W. B. Saunders Company, 1972.
331. North Atlantic Treaty Organization. Committee on Challenges of Modern Society. Air Pollution. Air Quality Criteria for Carbon Monoxide. N.10. Brussels: North Atlantic Treaty Organization, 1972. [265 pp.]
332. Noyes, W. A., Jr., and P. A. Leighton. Appendix CC-I,II, Photochemical reactions of carbonyls, p. 449. In The Photochemistry of Gases. New York: Dover Publications, Inc., 1966.

333. O'Donnell, R. D., P. Chikos, and J. Theodore. Effect of carbon monoxide exposure on human sleep and psychomotor performance. J. Appl. Physiol. 31:513–518, 1971.
334. O'Donnell, R. D., P. Mikulka, P. Heinig, and J. Theodore. Low level carbon monoxide exposure and human psychomotor performance. Toxicol. Appl. Pharmacol. 18:593–602, 1971.
335. O'Hanlon, J. F. Preliminary studies of the effects of carbon monoxide on vigilance in man, pp. 61–72. In B. Weiss and V. G. Laties, Eds. Behavioral Toxicology. New York: Plenum Press, 1975.
336. Ontario Department of Health. Perinatal Mortality Study Committee. Second Report of Perinatal Mortality Study in Ten University Teaching Hospitals, Ontario, Canada. Toronto: Ontario Department of Health, 1967. 274 pp.
337. Ontario Department of Health. Perinatal Mortality Study Committee. Supplement to the Second Report of the Perinatal Mortality Study in Ten University Teaching Hospitals, Ontario, Canada. Toronto: Ontario Department of Health, 1967, pp. 85–275.
338. Opitz, E. Increased vascularization of the tissue due to acclimatization to high altitude and its significance for oxygen transport. Exp. Med. Surg. 9:389–403, 1951.
339. Ott, W., J. H. Clark, and G. Ozolins. Calculating Future Carbon Monoxide Emissions and Concentrations from Urban Traffic Data. National Air Pollution Control Administration Publ. 999-AP-41. Durham, N.C.: U.S. Department of Health, Education, and Welfare, 1967. 40 pp.
340. Palanos, P. N. A practical design for an ambient carbon monoxide mercury replacement analyzer. Anal. Instrum. (Symposium) 10:117–125, 1972.
341. Palek, J., V. Brabec, L. Mirčevová, V. Volek, and B. Friedmann. Red cell carbohydrate metabolism in primary refractory anemias. Clin. Chim. Acta 18:39–45, 1967.
342. Park, C. D., L. Mela, R. Wharton, J. Reilly, P. Fishbein, and E. Aberdeen. Cardiac mitochondrial activity in acute and chronic cyanosis. J. Surg. Res. 14:139–146, 1973.
343. Partington, J. R. Carbon monoxide, pp. 686–696. In Textbook of Inorganic Chemistry for University Students. (5th ed.) London: MacMillan & Co., Limited, 1937.
344. Pauling, L. [Resonant energy of carbon monoxide], pp. 194–195. In The Nature of the Chemical Bond and the Structure of Molecules and Crystals: An Introduction to Modern Structural Chemistry. (3rd ed.) Ithaca, N.Y.: Cornell University Press, 1960.
345. Penney, D., M. Benjamin, and E. Dunham. Effect of carbon monoxide on cardiac weight as compared with altitude effects. J. Appl. Physiol. 37:80–84, 1974.
346. Penney, D., E. Dunham, and M. Benjamin. Chronic carbon monoxide exposure: Time course of hemoglobin, heart weight and lactate dehydrogenase isozyme changes. Toxicol. Appl. Pharmacol. 28:493–497, 1974.
347. Permutt, S., and L. Farhi. Tissue hypoxia and carbon monoxide, pp. 18–24. In National Research Council. Division of Medical Sciences. Committee on Effects of Atmospheric Contaminants on Human Health and Welfare. Effects of Chronic Exposure to Low Levels of Carbon Monoxide on Human Health, Behavior, and Performance. Washington, D.C.: National Academy of Sciences, 1969.
348. Pernoll, M. L., J. Metcalfe, T. L. Schlenker, J. E. Welch, and J. A. Matsumoto. Oxygen consumption at rest and during exercise in pregnancy. Respir. Physiol. 25: 285–293, 1975.
349. Perper, J. A., L. H. Kuller, and M. Cooper. Arteriosclerosis of coronary arteries in sudden, unexpected deaths. Circulation 52(6 Suppl. 3):27–33, 1975.
350. Peterson, W. F., K. N. Morese, and D. F. Kaltreider. Smoking and prematurity.

A preliminary report based on study of 7740 Caucasians. Obstet. Gynecol. 26:775–779, 1965.

351. Pierce, J. O., and R. J. Collins. Calibration of an infrared analyzer for continuous measurement of carbon monoxide. Amer. Ind. Hyg. Assoc. J. 32:457–462, 1971.
352. Pirnay, F., J. Dujardin, R. Deroanne, and J. M. Petit. Muscular exercise during intoxication by carbon monoxide. J. Appl. Physiol. 31:573–575, 1971.
353. Pitts, G. C., and N. Pace. The effect of blood carboxyhemoglobin concentration on hypoxia tolerance. Amer. J. Physiol. 148:139–151, 1947.
354. Plevová, J., and E. Frantík. The influence of various saturation rates on motor performance of rats exposed to carbon monoxide. Activ. Nerv. Super. 16:101–102, 1974.
355. Porter, K., and D. H. Volman, Flame ionization of carbon monoxide for gas chromatographic analysis. Anal. Chem. 34:748–749, 1962.
356. Power, G. G., and L. D. Longo. Fetal circulation times and their implications for tissue oxygenation. Gynecol. Invest. 6:342–355, 1975.
357. Pressman, J., L. M. Arin, and P. Warneck. Mechanisms for Removal of Carbon Monoxide from the Atmosphere. Report GCA TR-70-6-G. Bedford, Mass.: GCA Corporation, 1970. 45 pp.
358. Pressman, J., and P. Warneck. The stratosphere as a chemical sink for carbon monoxide. J. Atmos. Sci. 27:155–163, 1970.
359. Primary prevention of the atherosclerotic diseases, pp. 15–56. In I. S. Wright and D. T. Fredrickson, Eds. Cardiovascular Diseases. Guidelines for Prevention and Care. Reports of the Inter-Society Commission for Heart Disease Resources. Washinton, D.C.: U. S. Government Printing Office, 1973.
360. Quayle, J. R. The metabolism of one-carbon compounds by micro-organisms. Adv. Microb. Physiol. 7:119–203, 1972.
361. Radford, E. P., and E. Kresin. Red cell diphosphoglycerate after exposure to carbon monoxide. Fed. Proc. 32:350, 1973. (abstract)
362. Rahn, H., and W. O. Fenn. A Graphical Analysis of the Respiratory Gas Exchange. The O_2-CO_2 Diagram. Washington, D.C.: American Physiological Society, 1955. 38 pp.
363. Ramsey, J. M. Carbon monoxide, tissue hypoxia, and sensory psychomotor response in hypoxaemic subjects. Clin. Sci. 42:619–625, 1972.
364. Ramsey, J. M. Carboxyhemoglobinemia in parking garage employees. Arch. Environ. Health 15:580–583, 1967.
365. Ramsey, J. M. Effects of single exposures of carbon monoxide on sensory and psychomotor response. Amer. Ind. Hyg. J. 34:212–216, 1973.
366. Raven, P. B., B. L. Drinkwater, S. M. Horvath, R. O. Ruhling, J. A. Gliner, J. C. Sutton, and N. W. Bolduan. Age, smoking habits, heat stress, and their interactive effects with carbon monoxide and peroxyacetylnitrate on man's aerobic power. Int. J. Biometeorol. 18:222–232, 1974.
367. Raven, P. B., B. L. Drinkwater, R. O. Ruhling, N. Bolduan, S. Taguchi, J. Gliner, and S. M. Horvath. Effect of carbon monoxide and peroxyacetylnitrate on man's maximal aerobic capacity. J. Appl. Physiol. 36:288–293, 1974.
368. Ravenholt, R. T., M. J. Levinski, D. J. Nellist, and M. Takenaga. Effects of smoking upon reproduction. Amer. J. Obstet. Gynecol. 96:267–281, 1966.
369. Ray, A. M., and T. H. Rockwell. An exploratory study of automobile driving performance under the influence of low levels of carboxyhemoglobin. Ann. N.Y. Acad. Sci. 174:396–408, 1970.
370. Regan, D. Evoked Potentials in Psychology, Sensory Physiology and Clinical Medi-

cine. New York: John Wiley & Sons, Inc., 1972. 328 pp.

371. Rikans, L. E., and R. A. van Dyke. Evidence for a different CO-binding pigment in solubilized rat hepatic microsomes. Biochem. Pharmacol. 20:15–22, 1971.

372. Ringold, A., J. R. Goldsmith, H. L. Helwig, R. Finn, and F. Schuette. Estimating recent carbon monoxide exposures. A rapid method. Arch. Environ. Health 5:308–318, 1962.

373. Rissanen, V., K. Pyörälä, and O. P. Heinonen. Cigarette smoking in relation to coronary and aortic atherosclerosis. Acta Path. Microbiol. Scand. A 80:491–500, 1972.

374. Robbins, R. C., K. M. Borg, and E. Robinson. Carbon monoxide in the atmosphere. J. Air Pollut. Control Assoc. 18:106–110, 1968.

375. Roberts, W. C. Coronary arteries in fatal acute myocardial infarction. Circulation 45:215–230, 1972.

376. Robinson, E., and C. W. Moser. Global gaseous pollutant emissions and removal mechanism, pp. 1097–1101. In H. M. Englund and W. T. Beery, Eds. Proceedings of the Second International Clean Air Congress, Washington, D.C., 1970. New York: Academic Press, Inc., 1971.

377. Robinson, E., and R. C. Robbins. Atmospheric background concentrations of carbon monoxide. Ann. N.Y. Acad. Sci. 174:89–95, 1970.

378. Robinson, E., and R. C. Robbins. Sources, Abundance, and Fate of Gaseous Atmospheric Pollutants, SRI Project PR-6755. Huntsville, Ala.: Stanford Research Institute, 1968. 123 pp.

379. Rodkey, F. L. Kinetic aspects of cyanmethemoglobin formation from carboxyhemoglobin. Clin. Chem. 13:2–5, 1967.

380. Rodkey, F. L., J. D. O'Neal, H. A. Collison, and D. E. Uddin. Relative affinity of hemoglobin S and hemoglobin A for carbon monoxide and oxygen. Clin. Chem. 20:83–84, 1974.

381. Rondia, D. Abaissement de l'activite de la benzopyrène-hydroxylase hépatique *in vivo* après inhalation d'oxyde de carbone. C. R. Acad. Sci. D 271:617–619, 1970.

382. Roth, R. A., Jr., and R. J. Rubin. Comparison of the effect of carbon monoxide and of hypoxic hypoxia. I. *In vivo* metabolism, distribution and action of hexobarbital. J. Pharmacol. Exp. Ther. 199:53–60, 1976.

383. Roth, R. A., Jr., and R. J. Rubin. Comparison of the effect of carbon monoxide and of hypoxic hypoxia. I. *In vivo* metabolism, distribution and action of hexobarbital. J. Pharmacol. Exp. Ther. 199:53–60, 1976.

384. Roth, R. A., Jr., and R. J. Rubin. Role of blood flow in carbon monoxide- and hypoxic hypoxia-induced alterations in hexobarbital metabolism in rats. Drug. Metab. Dispos. 4:460–467, 1976.

385. Roth, R. P., R. T. Drew, R. J. Lo, and J. R. Fouts. Dichloromethane inhalation, carboxyhemoglobin concentrations, and drug metabolizing enzymes in rabbits. Toxicol. Appl. Pharmacol. 33:427–437, 1975.

386. Rottman, G. J., and H. W. Moos. The ultraviolet (1200–1900 angstrom) spectrum of Venus. J. Geophys. Res. 78:8033–8048, 1973.

387. Roughton, F. J. W., and R. C. Darling. The effect of carbon monoxide on the oxyhemoglobin dissociation curve. Amer. J. Physiol. 141:17–31, 1944.

388. Rummo, N., and K. Sarlanis. The effect of carbon monoxide on several measures of vigilance in a simulated driving task. J. Safety Res. 6:126–130, 1974.

389. Rush, D. Examination of the relationship between birthweight, cigarette smoking during pregnancy and maternal weight gain. J. Obstet. Gynaecol. Brit. Commonw. 81:746–752, 1974.

390. Russell, C. S., R. Taylor, and C. E. Law. Smoking in pregnancy, maternal blood pressure, pregnancy outcome, baby weight and growth, and related factors. A prospective study. Brit. J. Prev. Soc. Med. 22:119–126, 1968.
391. Russell, C. S., R. Taylor, and R. N. Maddison. Some effects of smoking in pregnancy. J. Obstet. Gynaecol. Brit. Commonw. 73:742–746, 1966.
392. Russell, M. A. H., P. V. Cole, C. Wilson, M. Idle, and C. Feyerabend. Comparison of increases in carboxyhemoglobin after smoking "extra-mild" and "non-mild" cigarettes. Lancet 2:687–690, 1973.
393. Sayers, R. R., W. P. Yant, E. Levy, and W. B. Fulton. Effect of Repeated Daily Exposure of Several Hours to Small Amounts of Automobile Exhaust Gas. Public Health Bulletin No. 186. Washington, D.C.: U.S. Government Printing Office, 1929. 58 pp.
394. Schoeneck, F. J. Cigarette smoking in pregnancy. N.Y. State J. Med. 41:1945–1948, 1941.
395. Scholander, P. F., and F. J. W. Roughton. Micro gasometric estimation of the blood gases. II. Carbon monoxide. J. Biol. Chem. 148:551–563, 1943.
396. Schulte, J. H. Effects of mild carbon monoxide intoxication. Arch. Environ. Health 7:524–530, 1963.
397. Seiler, W. The cycle of atmospheric CO. Tellus 26:116–135, 1974.
398. Seiler, W., and C. Junge. Carbon monoxide in the atmosphere. J. Geophys. Res. 75:2217–2226, 1970.
399. Seiler, W., and C. Junge. Decrease of carbon monoxide mixing ratio above the polar tropopause. Tellus 21:447–449, 1969.
400. Seiler, W., and P. Warneck. Decrease of carbon monoxide ratio at the tropopause. J. Geophys. Res. 77:3204–3214, 1972.
401. Shapiro, S., and J. Unger. Weight at Birth and Survival of the Newborn. United States, Early 1950. Vital and Health Statistics Series 21, No. 3. Washington, D.C.: U.S. Department of Health, Education, and Welfare, 1965. 33 pp.
402. Shaw, J. H. A Determination of the Abundance of Nitrous Oxide, Carbon Monoxide and Methane in Ground Level Air at Several Locations Near Columbus, Ohio. Scientific Report No. 1 on Contract AF19(604)2259. Air Force Cambridge Research Laboratory AFCRC-TN-59-428. Columbus: Ohio State University, Research Foundation, 1959. 44 pp.
403. Shaw, J. H. The abundance of atmospheric carbon monoxide above Columbus, Ohio. Astrophys. J. 128:428–440, 1958.
404. Shurtleff, D. Some characteristics related to the incidence of cardiovascular disease and death: Framingham Study, 16-year follow-up. Section 26. In W. B. Kannel and T. Gordon, Eds. The Framingham Study. An Epidemiological Investigation of Cardiovascular Disease. Washington, D.C.: U.S. Government Printing Office, December 1970.
405. Sidgwick, N. V. Carbon monoxide, pp. 545–546. In The Chemical Elements and Their Compounds. Vol. 1. Oxford, England: Clarendon Press, 1950.
406. Simonaitis, R., and J. Heicklen. Kinetics and mechanism of the reaction of $O(^3P)$ with carbon monoxide. J. Chem. Phys. 56:2004–2011, 1972.
407. Simpson, W. J. A preliminary report on cigarette smoking and the incidence of prematurity. Amer. J. Obstet. Gynecol. 73:808–815, 1957.
408. Sjöstrand, T. A method for the determination of carboxyhaemoglobin concentrations by analysis of the alveolar air. Acta Physiol. Scand. 16:201–210, 1948.
409. Small, K. A., E. P. Radford, J. M. Frazier, F. L. Rodkey, and H. A. Collison. A rapid method for simultaneous measurement of carboxy- and methemoglobin in

blood. J. Appl. Physiol. 31:154–160, 1971.

410. Smith, E., E. McMillan, and L. Mack. Factors influencing the lethal action of illuminating gas. J. Ind. Hyg. 17:18–20, 1935.
411. Smith, F., and A. C. Nelson, Jr. Guidelines for Development of a Quality Assurance Program. Reference Method for the Continuous Measurement of Carbon Monoxide in the Atmosphere. EPA-R4-73-028a. Research Triangle Park, N.C.: Research Triangle Institute, 1973. 110 pp.
412. Smith, K. A., J. M. Bremner, and M. A. Tabatabai. Sorption of gaseous atmospheric pollutants by soils. Soil Sci. 116:313–319, 1973.
413. Smith, L., Jr., and E. H. C. Sie. Response of luminescent bacteria to common atmospheric pollutants. Inst. Environ. Sci. 15:154–157, 1969.
414. Smith, R. N., and J. Mooi. The catalytic oxidation of carbon monoxide by nitrous oxide on carbon surfaces. J. Phys. Chem. 59:814–819, 1955.
415. Stainsby, W. N., and A. B. Otis. Blood flow, blood oxygen tension, oxygen uptake, and oxygen transport in skeletal muscle. Amer. J. Physiol. 206:858–866, 1964.
416. Stedman, D. H., E. D. Morris, Jr., E. E. Daby, H. Niki, and B. Weinstock. The role of OH radicals in photochemical smog reactions. Abstract WATR 026. In Abstracts of Papers. 160th National Meeting, American Chemical Society, Chicago, Illinois, September 14–18, 1970.
417. Stevens, C. M., L. Krout, D. Walling, A. Venters, A. Engelkemeir, and L. E. Ross. The isotopic composition of atmospheric carbon monoxide. Earth Planet. Sci. Lett. 16:147–165, 1972.
418. Stewart, R. D., E. D. Baretta, L. R. Platte, E. B. Stewart, H. C. Dodd, K. K. Donohoo, S. A. Graff, J. H. Kalbfleisch, C. L. Hake, B. van Yserloo, A. A. Rimm, and P. E. Newton. "Normal" Carboxyhemoglobin Levels of Blood Donors in the United States. Report No. ENVIR MED MCW CRC-COHb-73-1. Milwaukee: The Medical College of Wisconsin, 1973. [252 pp.]
419. Stewart, R. D., E. D. Baretta, L. R. Platte, E. B. Stewart, J. H. Kalbfleisch, B. van Yserloo, and A. A. Rimm. Carboxyhemoglobin concentrations in blood from donors in Chicago, Milwaukee, New York, and Los Angeles. Science 182:1362–1364, 1973.
420. Stewart, R. D., E. D. Baretta, L. R. Platte, E. B. Stewart, J. H. Kalbfleisch, B. van Yserloo, and A. A. Rimm. Carboxyhemoglobin levels in American blood donors. J.A.M.A. 229:1187–1195, 1974.
421. Stewart, R. D., T. N. Fisher, M. J. Hosko, J. E. Peterson, E. D. Baretta, and H. C. Dodd. Experimental human exposure to methylene chloride. Arch. Environ. Health 25:342–348, 1972.
422. Stewart, R. D., P. E. Newton, M. J. Hosko, and J. E. Peterson. Effect of carbon monoxide on time perception. Arch. Environ. Health 27:155–160, 1973.
423. Stewart, R. D., P. E. Newton, M. J. Hosko, J. E. Peterson, and J. W. Mellender. The effect of carbon monoxide on time perception, manual coordination, inspection, and arithmetic, pp. 29–55. In B. Weiss and V. G. Laties, Eds. Behavioral Toxicology. New York: Plenum Press, 1975.
424. Stewart, R. D., J. E. Peterson, E. D. Baretta, R. T. Bachand, M. J. Hosko, and A. A. Herrmann. Experimental human exposure to carbon monoxide. Arch. Environ. Health 21:154–164, 1970.
425. Stewart, R. D., J. E. Peterson, T. N. Fisher, M. J. Hosko, E. D. Baretta, H. C. Dodd, and A. A. Herrmann. Experimental human exposure to high concentrations of carbon monoxide. Arch. Environ. Health 26:1–7, 1973.
426. Strickland-Constable, R. F. Part played by surface oxides in the oxidation of carbon. Trans. Faraday Soc. 34:1074–1080, 1938.

427. Stupfel, M., and G. Bouley. Physiological and biochemical effects on rats and mice exposed to small concentrations of carbon monoxide for long periods. Ann. N.Y. Acad. Sci. 174:342–368, 1970.
428. Swinnerton, J. W., and R. A. Lamontagne. Carbon monoxide in the South Pacific Ocean. Tellus 26:136–142, 1974.
429. Swinnerton, J. W., V. J. Linnenbom, and C. H. Cheek. Distribution of methane and carbon monoxide between the atmosphere and natural waters. Environ. Sci. Technol. 3:836–838, 1969.
430. Swinnerton, J. W., V. J. Linnenbom, and R. A. Lamontagne. Distribution of carbon monoxide between the atmosphere and the ocean. Ann. N.Y. Acad. Sci. 174:96–101, 1970.
431. Swinnerton, J. W., V. J. Linnenbom, and R. A. Lamontagne. The ocean: A natural source of carbon monoxide. Science 167:984–986, 1970.
432. Syvertsen, G. R., and J. A. Harris. Erythropoietin production in dogs exposed to high altitude and carbon monoxide. Amer. J. Physiol. 225:293–299, 1973.
433. Tanaka, M. Studies on the etiological mechanism of fetal developmental disorders caused by maternal smoking during pregnancy. J. Jap. Obstet. Gynecol. Soc. (Nippon. Sanka-Fujinka Gakkai Zasshi) 17:1107–1114, 1965. (in Japanese)
434. Taylor, O. C. Air pollution with relation to agronomic crops. IV. Plant growth suppressed by exposure to air-borne oxidants (smog). Agron. J. 50:556–558, 1958.
435. Teichner, W. H. Carbon monoxide and human performance: A methodological exploration, pp. 77–103. In B. Weiss and V. G. Laties, Eds. Behavioral Toxicology. New York: Plenum Press, 1975.
436. Teichner, W. H. Recent studies of simple reaction time. Psychol. Bull. 51:128–149, 1954.
437. Teichner, W. H., and M. J. Krebs. Laws of the simple visual reaction time. Psychol. Rev. 79:344–358, 1972.
438. Tenney, S. M., and L. C. Ou. Physiological evidence for increased tissue capillarity in rats acclimatized to high altitude. Respir. Physiol. 8:137–150, 1970.
439. Theodore, J., R. D. O'Donnell, and K. C. Back. Toxicological evaluation of carbon monoxide in humans and other mammalian species. J. Occup. Med. 13:242–255, 1971.
440. Thomas Jefferson University. Jefferson Medical College, Department of Physiology. The Effect of Carbon Monoxide Inhalation on Induced Ventricular Fibrillation in the Cynomolgus Monkey. Technical Report. Philadelphia: Thomas Jefferson University, 1973. 31 pp.
441. Thompson, C. R., O. C. Taylor, M. D. Thomas, and J. O. Ivie. Effects of air pollutants on apparent photosynthesis and water use by citrus trees. Environ. Sci. Technol. 1:644–650, 1967.
442. Thomsen, H. K. Carbon monoxide-induced atherosclerosis in primates. An electron-microscopic study on the coronary arteries of *Macaca irus* monkeys. Atherosclerosis 20:233–240, 1974.
443. Thomsen, H. K., and K. Kjeldsen. Threshold limit for carbon monoxide-induced myocardial damage. An electron microscopic study in rabbits. Arch. Environ. Health 29:73–78, 1974.
444. Tjepkema, J. D., and C. S. Yocum. Leghemoglobin facilitated oxygen diffusion in soybean nodule slices. Plant Physiol. 46(Suppl.):44, 1970. (abstract)
445. Tokyo Metropolitan Research Institute for Environmental Protection. Annual Report of the Tokyo Metropolitan Research Institute for Environmental Protection. 1971. 68 pp.
446. Trapnell, B. M. W. Chemisorption. London: Butterworths Scientific Publications,

1955. 265 pp.

447. Tunder, R. S., S. W. Mayer, E. A. Cook, and L. Schieler. Compilation of Reaction Rate Data for Non-Equilibrium Performance and Reentry Calculation Programs. Aerospace Report No. TR-1001(9210-02)-1. El Segundo, Calif.: Aerospace Corporation, 1967. 68 pp.
448. Underwood, P. B., K. F. Kesler, J. M. O'Lane, and D. A. Callagan. Parental smoking empirically related to pregnancy outcome. Obstet. Gynecol. 29:1–8, 1967.
449. U.S. Bureau of the Census. Social and Economic Statistics Administration. Department of Commerce. Statistical Abstract of the United States 1974. 95th Annual Edition. Washington, D.C.: U.S. Government Printing Office, 1974. 1,028 pp.
450. U.S. Department of Health, Education, and Welfare. Public Health Service. Environmental Health Service. National Air Pollution Control Administration. Air Quality Criteria for Carbon Monoxide. NAPCA Publication No. AP-62. Washington, D.C.: U.S. Government Printing Office, 1970. [166 pp.]
451. U.S. Environmental Protection Agency. Ambient air monitoring reference and equivalent methods. Federal Register 40:7042–7070, 1975.
452. U.S. Environmental Protection Agency. Compilation of Air Pollutant Emission Factors (2nd ed., 3rd Printing with Supplements 1–5). AP-42. Research Triangle Park, N.C.: U.S. Environmental Protection Agency, 1976. [462 pp.]
453. U.S. Environmental Protection Agency. Guidelines for Air Quality Maintenance Planning and Analysis. Vol. 9. Evaluating Indirect Sources. EPA 450/4-75-001. Research Triangle Park, N.C.: U.S Environmental Protection Agency, Office of Air Quality, Planning and Standards, 1975. [557 pp.]
454. U.S. Environmental Protection Agency. National primary and secondary ambient air quality standards. Federal Register 36:8186–8201, 1971.
455. U.S. Environmental Protection Agency. Nationwide Emissions Report. United States (Emissions as of: March 12, 1975). National Emissions Data System (NEDS) of the Aerometric and Emissions Reporting System (AEROS). Research Triangle Park, N.C.: U.S. Environmental Protection Agency, 1975. (computer report available from agency)
456. U.S. Environmental Protection Agency. 1972 National Emissions Report: National Emissions Data System (NEDS) of the Aerometric and Emissions Reporting System (AEROS). EPA-450/2-74-012. Research Triangle Park, N.C.: U.S. Environmental Protection Agency, 1974. 422 pp.
457. van Kampen, E. J., and W. G. Zijlstra. Standardization of hemoglobinometry. II. The hemoglobincyanide method. Clin. Chim. Acta 6:538–544, 1961.
458. van Slyke, D. D., A. Hiller, J. R. Weisiger, and W. O. Cruz. Determination of carbon monoxide in blood and of total and active hemoglobin by carbon monoxide capacity. Inactive hemoglobin and methemoglobin contents of normal human blood. J. Biol. Chem. 166:121–148, 1946.
459. Vaughan, B. E., and N. Pace. Changes in myoglobin content of the high altitude acclimatized rat. Amer. J. Physiol. 185:549–556, 1956.
460. Vogel, J. A., and M. A. Gleser. Effect of carbon monoxide on oxygen transport during exercise. J. Appl. Physiol. 32:234–239, 1972.
461. Vogel, J. A., M. A. Gleser, R. C. Wheeler, and B. K. Whitten. Carbon monoxide and physical work capacity. Arch. Environ. Health 24:198–203, 1972.
462. Vollmer, E. P., G. B. King, J. E. Birren, and M. B. Fisher. The effects of carbon monoxide on three types of performance, at simulated altitudes of 10,000 and 15,000 feet. J. Exp. Psychol. 36:244–251, 1946.
463. Wald, N., S. Howard, P. G. Smith, and K. Kjeldsen. Association between atherosclerotic diseases and carboxyhaemoglobin levels in tobacco smokers. Brit. Med.

J. 1:761–765, 1973.

464. Wallace, N. D., G. L. Davis, R. B. Rutledge, and A. Kahn. Smoking and carboxyhemoglobin in St. Louis Metropolitan population. Theoretical and empirical considerations. Arch. Environ. Health 29:136–142, 1974.
465. Wanstrup, J., K. Kjeldsen, and P. Astrup. Acceleration of spontaneous intimal-subintimal changes in rabbit aorta by a prolonged moderate carbon monoxide exposure. Acta Path. Microbiol. Scand. 75:353–362, 1969.
466. Warneck, P. On the role of OH and HO_2 radicals in the troposphere. Tellus 26:39–46, 1974.
467. Weast, R. C., Ed. Handbook of Chemistry and Physics. (56th ed.) Cleveland, Ohio: The Chemical Rubber Company, 1975.
468. Webster, W. S., T. B. Clarkson, and H. B. Lofland. Carbon monoxide-aggravated atherosclerosis in the squirrel monkey. Exp. Molec. Path. 13:36–50, 1970.
469. Weinblatt, E., C. W. Frank, S. Shapiro, and R. V. Sager. Prognostic factors in angina pectoris—a prospective study. J. Chron. Dis. 21:231–245, 1968.
470. Weinblatt, E., S. Shapiro, C. W. Frank, and R. V. Sager. Prognosis of men after first myocardial infarction: Mortality and first recurrence in relation to selected parameters. Amer. J. Public Health 58:1329–1347, 1968.
471. Weinstock, B. Carbon monoxide: Residence time in the atmosphere. Science 166: 224–225, 1969.
472. Weinstock, B., and T. Y. Chang. The global balance of carbon monoxide. Tellus 26:108–115, 1974.
473. Weinstock, B., and H. Niki. Carbon monoxide balance in nature. Science 176:290–292, 1972.
474. Weir, F. W., and T. H. Rockwell. An Investigation of the Effects of Carbon Monoxide on Humans in the Driving Task. Final Report. Columbus: The Ohio State University Research Foundation, 1973. 170 pp.
475. Weiss, B., and V. G. Laties. Enhancement of human performance by caffeine and the amphetamines. Pharmacol. Rev. 14:1–36, 1962.
476. Weiss, H. R., and J. A. Cohen. Effects of low levels of carbon monoxide on rat brain and muscle tissue Po_2. Environ. Physiol. Biochem. 4:31–39, 1974.
477. Wells, L. L. The prenatal effect of carbon monoxide on albino rats and the resulting neuropathology. Biologist 15:80–81, 1933.
478. Went, F. W. On the nature of Aitken condensation nuclei. Tellus 18:549–556, 1966.
479. Went, F. W. Organic matter in the atmosphere and its possible relation to petroleum formation. Proc. Nat. Acad. Sci. U.S.A. 46:212–221, 1960.
480. Westberg, K., and N. Cohen. The Chemical Kinetics of Photochemical Smog as Analyzed by Computer. Aerospace Report No. ATR-70 (8107)-1. El Segundo, Calif.: Aerospace Corporation, 1969. 29 pp.
481. Westberg, K., N. Cohen, and K. W. Wilson. Carbon monoxide: Its role in photochemical smog formation. Science 171:1013–1015, 1971.
482. Westenberg, A. A. Carbon monoxide and nitric oxide consumption in polluted air. The carbon monoxide-hydroperoxyl reaction. Science 177:255–256, 1972.
483. Westenberg, A. A., and N. deHaas. Steady-state intermediate concentrations and rate constants. Some HO_2 results. J. Phys. Chem. 76:1586–1593, 1972.
484. Whalen, W. J. Intracellular PO_2 in heart and skeletal muscle. Physiologist 14:69–82, 1971.
485. White, J. J. Carbon monoxide and its relation to aircraft. U.S. Nav. Med. Bull. 30:151–165, 1932.
486. Wilks, S. S., J. F. Tomashefski, and R. T. Clark, Jr. Physiological effects of chronic

exposure to carbon monoxide. J. Appl. Physiol. 14:305–310, 1959.
487. Williams, I. R., and E. Smith. Blood picture, reproduction, and general condition during daily exposure to illuminating gas. Amer. J. Physiol. 110:611–615, 1935.
488. Wilmer, W. H. Effects of carbon monoxide upon the eye. Amer. J. Ophthalmol. 4:73–90, 1921.
489. Wilson, D. F., J. W. Swinnerton, and R. A. Lamontagne. Production of carbon monoxide and gaseous hydrocarbons in seawater: Relation to dissolved organic carbon. Science 168:1577–1579, 1970.
490. Windholz, M., Ed. Carbon monoxide, pp. 231–232. In The Merck Index. (9th ed.) Rahway, N.J.: Merck & Co., Inc., 1976.
491. Winneke, G. Behavioral effects of methylene chloride and carbon monoxide as assessed by sensory and psychomotor performance, pp. 130–144. In C. Xintaras, B. L. Johnson, and I. de Groot Eds. Behavioral Toxicology: Early Detection of Occupational Hazards. HEW Publ. No. (NIOSH) 74–126. Washington, D.C.: U.S. Department of Health, Education, and Welfare, 1974.
492. Wittenberg, J. B., C. A. Appleby, and B. A. Wittenberg. The kinetics of the reactions of leghemoglobin with oxygen and carbon monoxide. J. Biol. Chem. 247: 527–531, 1972.
493. Wofsy, S. C., J. C. McConnell, and M. D. McElroy. Atmospheric CH_4, CO and CO_2. J. Geophys. Res. 77:4477–4493, 1972.
494. Wohlrab, H., and G. B. Ogunmola. Carbon monoxide binding studies of cytochrome a_3 hemes in intact rat liver mitochondria. Biochemistry 10:1103–1106, 1971.
495. Wright, G., P. Randell, and R. J. Shephard. Carbon monoxide and driving skills. Arch. Environ. Health 27:349–354, 1973.
496. Xintaras, C., B. L. Johnson, C. E. Ulrich, R. E. Terrill, and M. F. Sobecki. Application of the evoked response technique in air pollution toxicology. Toxicol. Appl. Pharmacol. 8:77–87, 1966.
497. Xintaras, C., C. E. Ulrich, M. F. Sobecki, and R. E. Terrill. Brain potentials studied by computor analysis. Arch. Environ. Health 13:223–232, 1966.
498. Yamate, N., and A. Inoue. Continuous analyzer for carbon monoxide in ambient air by electrochemical technique. Kogai to Taisaku (J. Public Nuisance) 9:292–296, 1973. (in Japanese)
499. Yerushalmy, J. Mother's cigarette smoking and survival of infant. Amer. J. Obstet. Gynecol. 88:505–518, 1964.
500. Yerushalmy, J. The relationship of parents' cigarette smoking to outcome of pregnancy—implications as to the problem of inferring causation from observed associations. Amer. J. Epidemiol. 93:443–456, 1971.
501. Young, I. M., and L. G. C. Pugh. The carbon monoxide content of foetal and maternal blood. J. Obstet. Gynaecol. Brit. Commonw. 70:681–684, 1963.
502. Younoszai, M. K., and J. C. Haworth. Placental dimensions and relations in preterm, term and growth-retarded infants. Amer. J. Obstet. Gynecol. 103:265–271, 1969.
503. Younoszai, M. K., J. Peloso, and J. C. Haworth. Fetal growth retardation in rats exposed to cigarette smoke during pregnancy. Amer. J. Obstet. Gynecol. 104:1207–1213, 1969.
504. Zatsiorskiĭ, M., V. Kondrateev, and S. Solnishkova. Radiation of the flame of carbon monoxide and ozone and the mechanism of this reaction. Zh. Fiz. Khim. (Leningrad) 14:1521–1527, 1940. (in Russian)
505. Zimmerman, P. W., W. Crocker, and A. E. Hitchcock. Initiation and stimulation

of roots from exposure of plants to carbon monoxide. Contrib. Boyce Thompson Inst. 5:1–17, 1933.

506. Zimmerman, P. W., W. Crocker, and A. E. Hitchcock. The effect of carbon monoxide on plants. Contrib. Boyce Thompson Inst. 5:195–211, 1933.

507. Zorn, H. The partial oxygen pressure in the brain and liver at subtoxic concentrations of carbon monoxide. Staub-Reinhalt. Luft (Engl. Ed.) 32(4):24–29, 1972.

Index

Absorption spectrum of carbon monoxide, 5, 9, 16
Aerosol, 45
Agricultural burning, carbon monoxide from, 32, 33, 34
Air monitoring programs, 37-38
Air Quality Criteria for Carbon Monoxide, 2
Air quality models, 62-63
Aldehydes, photolysis of, 21
Altitude, carbon monoxide effects and, 157-60, 182
Alveolar gas, estimation of carboxyhemoglobin by measuring, 64, 202-4
Ambient carbon monoxide
  concentration, 52
    aerobic capacity and, 152-57
    and incidence of heart disease, 112, 113
    at high altitudes, 157-60
    temporal patterns in, 37
  exposure to, 71, 166-67
  myocardial infarction fatality and, 116-18
Analytic methods for determining carbon monoxide
  in air, 37-38
  in blood, 200-204
Angina pectoris
  cause of, 109
  exercise and, 122, 123
  relation between carbon monoxide and, 120-22
  smoking and, 119, 124-25
Animals
  effect of carbon monoxide on
    behavioral vigilance, 136
    brain and liver, 80
    fetal and maternal carboxyhemoglobin concentration, 85-86, 100-101
    fetal growth, 93-94
    newborn, 95-96
  effect of chronic carbon monoxide exposure on, 163-64
  study of coronary atherosclerosis in, 110, 111
  study of myocardial toxicity in, 107-8
  suggested research on carbon monoxide effects on, 190
  tolerance to carbon monoxide exposure, 160-61
Anthropogenic sources of carbon monoxide
  amount generated, 30, 52

estimated emission from, 36–37, 47
from agricultural burning, 32, 33
from industrial processes, 31
from transportation, 30
Arteriosclerotic heart disease (ASHD), carbon monoxide exposure and, 108–9, 112
Atherosclerosis. *See* Coronary atherosclerosis
Atmosphere, carbon monoxide concentration in, 29–30
Atmospheric reactions of carbon monoxide, 16
chemical modeling of, 23–37
with atomic oxygen, 18–20
with hydroxyl radicals, 20–21
with molecular species, 17
with trace contaminants, 22–23
Automated carbon monoxide determination, 193
performance specifications for, 194
specifications for interfering substances, 195
Automobiles. *See* Motor vehicles
Avian embryogenesis, carbon monoxide effects on, 94–95

Bacteria
carbon monoxide effects on, 168–69, 184–85
oxidation of carbon monoxide by, 42–43
Behavioral effects of carbon monoxide, 127, 180–81
complex intellectual behavior, 150–51
coordination and tracking, 147–48
driving, 136–40
reaction time, 140–41
sensory processes, 148–50
time discrimination and estimation, 141–47
vigilance, 127, 130–36
Biologic effects of carbon monoxide, 2, 176
high altitude and, 103–4
maternal smoking and, 90–93
on avian embryogenesis, 94–95
on fetal carboxyhemoglobin, 84–90
on fetal growth, 93–94, 96–103
on maternal carboxyhemoglobin, 83, 85–89
on newborn, 95–103
Blood. *See also* Carboxyhemoglobin
carbon monoxide bound to hemoglobin in, 72, 74
carbon monoxide uptake by, 44
impairment of oxygen transport by, 74–79
oxyhemoglobin concentration of, 69, 158
oxyhemoglobin dissociation curve of, 74–76, 104, 108, 158, 177
procedures for determining carbon monoxide in
spectrophotometric, 200–201
volumetric, 201–2
Bond strength of carbon monoxide, 6
Brain, effect of carbon monoxide inhalation on, 79, 80

Carbon monoxide. *See also* Ambient carbon monoxide concentration; Behavioral effects of carbon monoxide; Biologic effects of carbon monoxide; Chemical reactions of carbon monoxide; Concentration of carbon monoxide; Emission of carbon monoxide; Exposure to carbon monoxide; Intracellular effects of carbon monoxide; Monitoring carbon monoxide; Natural sources of carbon monoxide; Production of carbon monoxide
chemical properties, 5
isotopic types of, 35
laboratory preparation of, 8
measurement of, 12–13, 61–62
residence times for, 38–40
solubility of, 15
Carbon monoxide poisoning
causes of, 1
compared with hypoxia, 82
history of, 1–2
symptoms, 106
venous blood oxygen partial pressure in, 74–77
Carbon monoxide uptake
by blood hemoglobin, 44
by plants, 43–44
by soil, 42
carboxyhemoglobin to determine, 16
endogenous versus exogenous, 68, 176
physiologic factors determining, 68–72
problems in calculating, 62

Carbon oxides, 4-5
Carbonyl compounds, 13-14
Carboxyhemoglobin. *See also* Blood
  calculation of, as function of time, 69
  concentration
    aerobic capacity and, 152-57
    altitude tolerance and, 159
    behavioral vigilance and, 133-36
    coronary blood flow and, 107-8
    dose-response relationship at, 165-66
    driving performance and, 137-39
    effect on intellectual behavior, 151
    effect on metabolism, 2, 42-43
    effect on sensory processes, 149-50
    effect on time discrimination and estimation, 145-46
    in mother and fetus, 83-93
    in postmortem ASHD deaths, 112, 114
    in smokers versus nonsmokers, 64-67, 70-73, 125-26
    reaction time and, 140-41
  formation of, 16
  methods of measuring, 175
    blood samples for, 64-66, 199-202
    gas chromatography, 66, 196-97
    measuring alveolar gas, 64, 202-4
    spectrophotometric, 200-201
    volumetric, 201-2
  relationship between cause of death and, 114-16
Carbureted water gas, 1, 7
Cardiovascular system
    effect of carbon monoxide on, 72-79
    response to carbon monoxide exposure, 104-08
Catalytic combustion-thermal detection, 194, 196
Catalytic converter, effect on carbon monoxide emission, 30, 51
Chemical reactions of carbon monoxide, 172-73
  combustion, 11
  decomposition, 10
  gas-phase, 18-23
  heterogeneous, 12-13
  models of, 23-27
  to form carbonyl compounds, 13-14
  to form coordination compounds, 14-16
  with molecular species, 17
Chemisorption of carbon monoxide, 45
Chlorosulfonic acid, 8-9
Cholesterol, 110
Claudication, effect of carbon monoxide exposure on, 122, 179
Combustion, 11
Concentration of carbon monoxide
  alveolar, 70
  atmospheric, 29-30, 40, 175
  distribution of, 39, 53
  indoor-outdoor differences in, 59, 60
  meteorological effects of, 59
  nonurban, 38
  range of, 38
  spatial variations in, 56-58
  temperature variations in, 54-56, 60
  urban, 37-38, 52-57, 59, 174, 175
Coordination ability, carbon monoxide exposure and, 147
Coordination compounds, 14-16
Coronary arterial stenosis, 109, 114
Coronary atherosclerosis, 109
  carbon monoxide exposure and, 110-11
  suggested studies on, 187
Cytochrome P-450 system, effect of carbon monoxide on, 81-82

Decomposition of carbon monoxide, 10
Dose-response relationships, of carbon monoxide, 165-66, 182-84, 187
Driving performance, carbon monoxide exposure and, 136-40, 189
Dual-isotope fluorescence, 194

Electric dipole moment, of carbon monoxide, 5
Electrochemical method for measuring carbon monoxide, 196
Elimination of carbon monoxide, 68-71
Emission of carbon monoxide
  conditions influencing, 50
  estimates of, 34, 46-47, 52
  from agricultural burning, 32, 33, 34
  from industrial processes, 31, 33
  from internal-combustion engine, 1-2, 30
  from motor vehicles, 30, 48-49, 51
  global, 32, 36-37, 173
  natural, 34-36
  urban, 48, 52-57
Environmental carbon monoxide. *See* Concentration of carbon monoxide

Exercise, effect of carbon monoxide on, 152-57, 181
angina pectoris and, 122, 123
suggested studies on, 186, 187
Exposure to carbon monoxide
adaptation to, 160-62
behavioral effects from, 127, 130-51
chronic, 162-64, 182
dose-response relationship to, 165-66, 182-84
effects of, 4, 47, 152-157
fetal, 90-93, 96, 166
from internal-combustion engine, 1-2, 30, 31
from smoking, 70-71, 72, 73
heart disease and, 108-9
incidence of, 112-14
severity of, 110
measuring period for, 47
morbidity due to heart disease and, 119-24
problems in calculating, 62-63

Fermentation, carbon monoxide in, 42-43
Fetus, effect of carbon monoxide on, 83-93, 177-78, 188
Fires, carbon monoxide production from, 11, 33, 34
Forest fires, as natural source of carbon monoxide, 34
Formaldehyde
as natural source of carbon monoxide, 35
photolysis of, 21, 26
Formic acid, 8-9, 12
Fossil fuels, carbon monoxide from combustion of, 33
Frontal analysis, to monitor carbon monoxide, 197
Fungi
oxidation of carbon monoxide by, 169
removal of carbon monoxide from atmosphere by, 42

Gas chromatography
to detect carbon monoxide in ambient air, 12
to measure carboxyhemoglobin, 66
Gas chromatography-flame ionization, 196-97
Gasoline, carbon monoxide from combustion of, 30

Health effects of carbon monoxide, 2, 5, 46, 164, 175
Heart
carbon monoxide and disease of, 108-26, 128, 129
oxygen carrying capacity of myocardium, 104-5, 108
response to carbon monoxide exposure, 104-8
Hemoglobin. *See* Blood; Carboxyhemoglobin
Hydrogen, competition between carbon monoxide and, 23
Hydroxyl radicals, processes generating, 20-23, 172-73
Hypoxemia, 81, 158
Hypoxia
carbon monoxide versus high altitude, 103-4, 158
compared with carbon monoxide poisoning, 86
effect of tissue oxygenation, 103
myocardial adaptation to, 104

Illuminating gas. *See* Water gas
Industrial sources of carbon monoxide, 11, 31, 33, 34
suggested methods for measuring, 188
Infrared region, carbon monoxide absorption of radiation in, 5, 191
Inhalation of carbon monoxide, 1, 68
accidental death from, 4
cardiovascular response to, 106
effect on brain and liver, 80
Intellectual behavior, carbon monoxide exposure and, 150-51
Interatomic distance, of carbon monoxide, 5
Internal-combustion engine, carbon monoxide emission from, 1-2, 30, 31
Intracellular effects of carbon monoxide
on cytochrome P-450 system, 81-82
on mitochondria, 79-80
on myoglobin, 79, 82-83
suggested studies on, 187
Iodometry, 12

Iron pentacarbonyl, reaction with carbon monoxide, 14-15
Isotopic types of carbon monoxide, 35

Liver, effect of carbon monoxide inhalation on, 80
Lungs
carboxyhemoglobin in chronic disease of, 71-72
measuring alveolar gas of, 64, 202

Measurement of carbon monoxide, 12-13, 61-62. *See also* Monitoring carbon monoxide
Mercury replacement, to monitor carbon monoxide, 197-98
Metabolism, effect of carbon monoxide on, 2, 42-43
Methane
as natural source of carbon monoxide, 35
carbon monoxide from oxidation of, 28, 173
conversion of carbon monoxide to, 12
Mitochondria, carbon monoxide and, 79-80, 83
Models
air quality, 62-63
for carbon monoxide reactions, 23-27
Monitoring carbon monoxide, 63-67, 174-75
methods for
analysis of blood samples, 199-202
catalytic combustion-thermal detection, 194, 196
dual-isotope fluorescence, 193
electrochemical, 196
frontal analysis, 197
gas chromatography-flame ionization, 196-97
measurement of alveolar gas, 199, 202-4
mercury replacement, 197-98
nondispersive infrared spectrometry, 191-93
suggested improvements in, 187
Motor vehicles, carbon monoxide emission from, 30, 48-49, 51
Myocardial infarction, 109, 112
ambient carbon monoxide and fatality from, 116-18
effect of smoking on risk of death from, 124
Myoglobin, carbon monoxide and, 79, 82-83

Natural gases, as source of carbon monoxide, 35
Natural sources of carbon monoxide, 29
estimated emission from, 36-37, 47
from forest fires, 34-35
from methane and formaldehyde, 35
from oceans, 36
of biologic or geophysical origin, 35
suggested studies for measuring, 190
Nickel carbonyl, carbon monoxide from decomposition of, 9
Nitrogen, similarity to carbon monoxide, 6
Nitrogen dioxide
carbon monoxide reaction with, 17, 18
conversion of nitric oxide to, 22, 24
Nitrous oxide, reaction with carbon monoxide, 45
Nondispersive infrared spectrometry, 191-93

Occupational source of carbon monoxide, 71, 124
Oceans
as sink for carbon monoxide, 44
as source of carbon monoxide, 36
Oxidation of carbon monoxide, 13, 16-17, 23
Oxygen
carbon monoxide effects on fetal blood, 96-103
combustion from carbon monoxide reaction with, 11
effect of carbon monoxide on transport system of, 72, 74-76
oxidation of carbon monoxide in, 13
reaction between carbon monoxide and atmospheric atomic, 18-20
Oxyhemoglobin
at high altitudes, 158
concentration, 69
dissociation, 74-76, 104, 108, 158, 177
Ozone, reaction with carbon monoxide, 17, 18, 24

Particles, as sinks for atmospheric carbon

monoxide, 45
Photochemical reactions of carbon monoxide, 10, 25-27, 36
Photochemical smog
conversion of nitric oxide to nitrogen dioxide in, 22, 23
effect of carbon monoxide on, 25
Photochemistry of carbon monoxide, 5
Photolysis
of aldehydes, 21
of carbon monoxide, 9, 10
of formaldehyde, 21
of hydrogen peroxide, 23
Physical properties of carbon monoxide, 5-6, 7
Physiologic effects of carbon monoxide
on respiratory and cardiovascular systems, 72-79
on tissues, 79-83
Plants
as sinks for carbon monoxide, 43-44
carbon monoxide effects on, 169-71, 184-85
Production of carbon monoxide, 6-9
anthropogenic, 28-29, 30, 31-33
by atmospheric photochemical oxidation, 35, 36
in soils, 42
natural, 29, 34-36
rates of, 37

Reaction time, carbon monoxide exposure and, 140-41
Recommended studies on effects of carbon monoxide on man and animals, 186-90
Reproductive system, carbon monoxide effects on, 83-90
fetal growth, 93-94
from maternal smoking, 90-93
Residence time, for carbon monoxide, 39-40
Respiratory system, carbon monoxide effects on, 68-79. *See also* Inhalation of carbon monoxide

Samples, carbon monoxide
from air, 37-38
from blood, 64-66, 199-202

Sensory processes, carbon monoxide exposure and, 148-50
Sinks for carbon monoxide, 40, 175
common surfaces as, 45
in blood hemoglobin, 44
in oceans, 44
in plants, 43-44
in soil, 41-43
in upper atmosphere, 41
suggested methods for measuring, 190
Skeletal muscle, toxic effects of carbon monoxide on, 82-83
Smoking, 3
aerobic capacity and, 152
carboxyhemoglobin concentration and, 64-67, 70-73, 125-26
effect on children from maternal, 92-93
effect on fetus, 178
carboxyhemoglobin concentration, 84-91
growth, 93-94
low birthweight, 91-92
oxygenation, 102-3
effect on mother
carboxyhemoglobin concentration, 83, 84-91
complications of pregnancy and labor, 92
heart disease and, 114, 123, 128-29, 179-80
morbidity from heart disease and, 119, 123
recommendations for publicizing hazardous effects of, 186
recommended research on role of carbon monoxide in, 186
Soils
as sink for carbon monoxide, 41-42
conversion of carbon monoxide in, 169
Solubility of carbon monoxide, 15
Sources of carbon monoxide, 28-29, 52, 173-74
anthropogenic, 30, 31, 32
natural, 29, 34-36
technological, 31-33
Spectrophotometric methods, 200-201
Spectroscopy, 45
Stoichiometric mixtures
of carbon monoxide and oxygen, 11
of carbon monoxide and water, 12

Storms, as source of carbon monoxide, 35
Stratosphere, carbon monoxide concentration in, 39, 41
Susceptibility to carbon monoxide
  at high altitudes, 157-60
  by plants, 170
  suggested studies on, 188
Symptoms
  of carbon monoxide exposure, 163-64
  of carbon monoxide toxicity, 106

Technological sources of carbon monoxide, 31-33
Threshold
  of carbon monoxide concentration, 46
  of carboxyhemoglobin concentration, 146-47
Time discrimination, carbon monoxide exposure and, 141-47
Tolerance to carbon monoxide exposure, 160-61
Toxicity of carbon monoxide, 16, 106
Trace contaminants, reaction with carbon monoxide, 16-17
Tracking, carbon monoxide exposure and, 147-48
Transportation, as source of carbon monoxide, 30, 34. *See also* Motor Vehicles
Troposphere. carbon monoxide concentration in, 40, 41

Ultraviolet region, carbon monoxide absorption of radiation in, 5, 9
Uptake. *See* Carbon monoxide uptake
Urban areas
  carbon monoxide concentration in, 37-38, 52-57, 174, 175
  carbon monoxide emission in, 48, 52-57

Venous blood oxygen partial pressure, 74-77
Vigilance, effect of carbon monoxide on
  auditory, 127, 130-31, 135
  nervous system activities, 136
  suggested studies on, 186-87
  visual, 131-34
Volcanoes, as source of carbon monoxide, 35
Volumetric methods, for measuring carbon monoxide, 201-2

Water gas, 1, 6-8
Water vapor, carbon monoxide reaction with, 17